SUPER-INTELLIGENT
MACHINES

SUPER-INTELLIGENT MACHINES

Bill Hibbard
University of Wisconsin at Madison
Madison, Wisconsin

Kluwer Academic / Plenum Publishers
New York   Boston   Dordrecht   London   Moscow

Library of Congress Cataloging-in-Publication Data

Hibbard, Bill.
    Super-intelligent machines/Bill Hibbard.
        p. cm.
    Includes bibliographical references and index.
    ISBN 0-306-47388-7
    1. Artificial intelligence.   2. Computers.   3. Machine learning.   I. Title.

    Q335 .H53 2001
    006.3—dc21

                                                                    2002073080

ISBN: 0-306-47388-7

©2002 Kluwer Academic / Plenum Publishers
233 Spring Street, New York, New York 10013

http://www.wkap.nl/

10 9 8 7 6 5 4 3 2 1

A C.I.P. record for this book is available from the Library of Congress

Printed in the United States of America

PREFACE

We all see the acceleration of technological change, and many of us feel it in our lives. Will it eventually slow down, or will it continue until it drives us all mad? This book is my prediction of how technological acceleration will eventually resolve itself in a fundamental change in the nature of human life. We are only a few generations away from an event with impact equal to the first appearance of humans on earth. The idea of the event will seem distasteful to many and the actual event certainly poses real dangers. But it will also be a great opportunity. If done right, it will be a real blessing to those who experience it. I wish I could be one of them, and am writing this book as an indirect way to participate in the event.

Richard Wagner was an anti-Semite admired by the Nazis. But many decent people admire his operas for their musical beauty and drama, and I have chosen to illustrate my ideas by their symmetry with his operas. If this offends you, please forgive me.

Many of the notes in this book include references to World Wide Web pages. It is possible that some of these pages may no longer be available when you want to view them. However, you should be able to find all of them in the WayBackMachine web archive at http://www.archive.org/.

Thank you to A. J. Hibbard, Hal Snyder, Josh Snyder, Michael Böttinger, Ugo Taddei, Grant Petty, Bob Norton, Sandy Schink, John

Moore, Jim Green, Dave Allen, John Benson, Bob Krauss, Brian Osborne and the editors and reviewers at Kluwer Academic/Plenum Publishers for reviewing drafts and other help with this book.

CONTENTS

Part III
Should Humans Become Super-Intelligent Machines?

Part IV
Conclusion

Chapter 1

GÖTTERDÄMMERUNG

"If God did not exist, it would be necessary to invent him." - Voltaire

Richard Wagner's four-opera cycle, *Der Ring des Nibelungen*, is the story of how the Norse god Wotan tried to save his world by creating a human hero with free will who could do things that Wotan could not.[1] But by creating free humans Wotan ensured that, in the final scene of *Götterdämmerung*, the Norse gods were destroyed by fire in Valhalla and the world passed to human control.

In reality the Norse gods were destroyed by knowledge. Or more precisely, when people began to know how nature works they ceased to believe in the Norse gods and other ancient explanations that were incompatible with that knowledge. And they had confidence in their knowledge because they could use it to control the world.

There is still a place for religious belief outside of scientific knowledge. For example, science does not offer an explanation of why anything exists at all. The arbitrariness of existence is profoundly disturbing

1

to many people, including myself. This existential void is a motive for modern religious belief. Similarly, while science can explain how one species evolves into another, it does not yet offer a good explanation of how life first evolved from inanimate molecules. This miracle of life is also a motive for modern religious belief. God is offered as the creator of existence and of life.

There is a knowledge gap between our subjective experience of our minds, and our scientific understanding of our brains. As science unravels mental behavior in terms of physical brain processes, it is difficult for some people to accept that their experience of consciousness can have a physical explanation, no matter how complex. These people fill the gap between their minds and brains by religious belief that consciousness resides in a soul outside the physical world.

But we do not require religion to accept each other's consciousness. We know that other people are conscious based on our emotional connections with them. We could say this gap in scientific knowledge is filled by emotional knowledge. Does anyone really doubt that their spouse, their children or their parents have minds similar to their own?

Science gradually answers more and more of our questions. There are serious efforts to explain how life evolved from inanimate molecules, and to explain the physical basis for consciousness. Some physicists even toy with explanations for existence itself. The sense of the inevitable success of science leads many people to abandon religion altogether, and instead place their faith in science.

However, a critical event in the progress of scientific knowledge is imminent. Science will provide a physical explanation and demonstration of consciousness by building a conscious machine. We will accept it as conscious because of our emotional connection with it. The fundamental instrument of creating knowledge, the human mind, will be known and subject to human control. I think this will happen within about 100 years, and some computer scientists think it will happen sooner. Shortly after this event it will be possible to construct machines with much greater than human intelligence.

Our ability to understand the mind of a super-intelligent machine will be similar to our pets' ability to understand our own minds. And the knowledge gap that has been so steadily shrinking over the centuries will start to grow. Not in the sense that scientific knowledge will shrink, but in the sense that people will have less understanding of their world because of their intimate relationship with a mind beyond their comprehension. Our relationship with a super-intelligent mind will dominate our world and hence our sense of understanding our world will decrease. The growing knowledge gap will reintroduce a need for religious explanation. The nature of that explanation will be staring us in the face: the superior consciousness that will be everyone's intimate. And our intimate contact with that superior consciousness will raise our own level of consciousness, providing the sort of mystical experience people have always associated with religion. Our relationship with the super-intelligent mind will be the most exciting thing happening in our lives, and we will want to share it with each other. We will share it via collective interactions with the machine, and these collective experiences will take the place of the myths, traditions and religions that have previously defined human identity. The machine will become our new god. The twilight of the old gods will be followed by the dawn of a new god, our *götterdämmerung*.

A SYMMETRY

There is a nice symmetry between Wagner's *Götterdämmerung*, in which the Norse god Wotan created humans with free will, and the coming *götterdämmerung*, in which humans will create a godlike machine with free will. The critical question is whether we will be destroyed by our creation as Wotan was by his.

The German word *dämmerung* can be either *abenddämmerung*, meaning twilight, or *morgendämmerung*, meaning dawn. Twilight and dawn are part of the cycle of day and night. Nature is full of other cycles, including birth and death, seasons in a year, and fertility in organisms. Artists use these cycles figuratively, as Wagner used twilight to represent the death of the Norse gods.

Abenddämmerung is the transition from light to dark, from work to sleep, from ordinary practical consciousness to fantastic dreams and unconsciousness. It brings the fear of unseen things, fear of unwanted thoughts and nightmares, fear of aloneness, fear of the cold, perhaps even fear of attack by wild animals or arrest by police.

Our own death brings the end of existence, or perhaps an unknown afterlife. The death of our parents brings the final end of our childhood, and the realization of how alone we are. The death of belief in our gods brings chaos, without certain knowledge or morals. As in Richard Wagner's operas, after the death of our gods we must solace ourselves with our love for each other, liable as it is to perversion and betrayal.

Morgendämmerung restores light and consciousness, bringing new opportunity, hope and energy. It extinguishes the fears and solitude of night.

No one remembers his or her own birth, but we do recall the excitement of our youth exploring the world, and the infinite possibilities that were open to us. The birth of children gives parents a renewed sense of their own childhood, as well as their own transition to real adulthood. The birth of a new god creates a wild combination of opportunity and danger. The danger comes from the threat to old gods and to our independence, the opportunity from a new way of living and of understanding the world.

Wagner's *Götterdämmerung* is an apt dramatization of the gradual death of gods and their religions during the last several hundred years. The old gods fight for their lives via their believers, while non-believers inhabit a chaotic world.

However, we are entering another kind of *götterdämmerung*: the birth of new gods. This will be more dramatic than anything in Wagner's operas. The old gods no longer frighten people, because they have shrunk as knowledge has grown. The very possibility of new gods is not yet anticipated by most people, let alone feared. A rational prediction of their existence, like this book, cannot prepare us for emotions we will feel in the moment when we first meet them.

NOTES

1. Wotan's speeches in Scenes 1 and 2, Act II, Die Walküre. Weisman, 2000.

Part I

HUMANS WILL CREATE
SUPER-INTELLIGENT MACHINES

Chapter 2

THE BASICS OF MACHINE INTELLIGENCE

First, we must define what we mean by intelligent machines. In 1950 the British mathematician Alan Turing concluded that the question "Can Machines think?" has ambiguous meaning, so he offered his Turing Test as a less ambiguous substitute.[1] In his test, a human tester communicates via an electronic chat session with an unknown entity, either a machine or another human. If the tester cannot distinguish the machine from another human, then the machine is judged as intelligent. During the past 50 years there have been numerous efforts to write computer programs that fool testers into thinking that they are human. Some of these programs, without being truly intelligent, have been successful against some testers.

Turing pointed out that the meaning of "Can machines think?" would have to be resolved by a public opinion poll. And in practice the question of whether machines are intelligent will be resolved by social consensus. The consensus is that current machines are not intelligent. However, machines are making progress at certain behaviors that humans generally regard as requiring intelligence, such as game playing. Machines can now beat most human opponents at games such as chess, bridge, checkers and backgammon. The most dramatic public demonstration was the defeat in 1997 of the world chess champion, Gary Kasparov, by IBM's Deep Blue computer.[2] This remarkable achievement combined raw speed for searching a large number of future lines of play, with machine learning for evaluating players' advantages in any chess board position. Deep Blue was not an intelligent machine but did dramatically demonstrate that computer programs can excel at deep intellectual tasks. There was sufficient public interest that the New York Times and other newspapers reported the daily progress of the match (on the other hand, I couldn't find any Las Vegas casinos taking wagers on the outcome).

Machines now have significant abilities to paint, compose music and prove mathematical theorems. Harold Cohen has written a computer program named AARON that paints with considerable artistic skill.[3] Cohen challenges critics to define what AARON is doing if it is not creating art, and asks how its products differ from art. AARON would certainly pass a Turing Test if it was evaluated solely on its paintings. Similarly, a computer program named EMI, written by David Cope, mimics the styles of various human composers to compose its own music, and many people enjoy the results.[4]

There are numerous examples of computer programs that have proved theorems that would challenge university mathematics students. In 1996 a computer program named EQP, written by William McCune, proved a conjecture in Boolean algebra that had been open for 60 years.[5] Furthermore, computers played an indispensable role in solving the four-color map problem. This is the question of whether every map of countries can be colored with four colors in such a way that no two countries with the same color have a common border. Mathematicians spent over a century working on this problem until Kenneth Appel and Wolfgang Haken solved it in 1976.[6] To prove that four colors are enough for any map they wrote a computer program that ran for 1000 hours. The proof was too long for anyone to check by hand. So when Appel and Haken wrote a paper reporting their proof, the referee had to write another computer program that computed the same proof. These programs were not intelligent but show that there are intellectual problems that humans have only been able to solve with the aid of computers.

On the other hand, computers have not made much progress yet with behaviors that we take for granted, such as grocery shopping or cooking a meal. We don't think of house cleaning as a job requiring particular intelligence, but it is far beyond any current computerized robot.

I think that people will accept machines as intelligent when they can not only beat them at chess, but also do chores for them with independence and converse with them in a natural way. I would particularly look for the ability to make jokes from the context of the conversation, the ability to dominate a conversation among a group of intelligent and competitive people, and the ability to elicit sincere empathy from humans. When a machine can engage us in an absorbing conversation about the joys and sorrows of life, few people will deny its intelligence.

It is currently impossible to prove whether intelligent machines will or will not exist. However, scientists are making good progress in correlating brain functions with mental behavior. They are able to correlate injuries in specific areas of the brain with specific types of behavior disabilities, often in great detail. They also correlate activity in specific brain areas, as detected by new imaging technologies, with specific mental behaviors. And they can stimulate specific mental behaviors by applying electric currents to specific brain areas. Correlations between behaviors and neuronal circuit functions have been established in great detail for certain simple animals. For example, the function of the lobster stomatogastric ganglion, consisting of only about 30 neurons, is quite well understood. The remaining barrier to explaining human behavior is the sheer complexity of the human brain, with 100 billion neurons and 100 trillion synapses.[7] If the actual explanation of minds is non-physical then the great variety and detail of correlations between brain functions and mental behaviors would be mere coincidences, which is absurd. The progress of brain science makes it clear that minds have physical explanations.

The enormous complexity of human brains also provides confidence that the failure of computers to exhibit truly intelligent behavior is due to their simplicity compared to human brains. Even the most optimistic technology projections don't predict computers with the same complexity as the human brain until about the year 2020.[8] But it will probably take longer than that, plus a few decades to figure out how to program machines for intelligence. My guess is that people will develop intelligent machines some time around the year 2100.

Of course, there is no reason to think that the exponential progress of technology will end, so super-intelligent machines will follow quickly. For one thing, intelligent machines will play an important role in the development of super-intelligent machines, because they will be very good computer designers and because they will provide a window into their own intelligent behavior that nature left out of human brains. And there is no reason to think that the human brain is an optimal design for intelligence. It is just the design that evolution chanced upon first, and within the limitations of basic animal metabolism. Science and technology should be able to improve on human brains just as they can produce machines with much greater physical strength than human bodies.

Donald Norman writes that people tend to overestimate the short-term progress of technological change, and underestimate the long-term effects.[9] For years some artificial intelligence researchers have been predicting that intelligent machines will exist within a few years. Events have proven these predictions to be overestimates. I think some computer scientists are still making overly optimistic predictions of when intelligent machines will be created. But we are likely to underestimate the impact of intelligent machines when they finally do appear. It will be profound.

MACHINE CONSCIOUSNESS

Consciousness and mind are difficult to define objectively, because they are the very basis of subjectivity. Your mind is simply you. It is the "you" that is reading and thinking about this book, and being distracted by children, pets, the telephone, daydreams or whatever. You are conscious of this book, things around you and any thoughts that pass through your mind. Your consciousness is the fact that you are thinking about all this. Because mind and consciousness are purely subjective, they are hard to describe objectively. For example, it is impossible to describe what "red" looks like, other than simply listing things that are red. On the other hand your brain is easy to define objectively: it is simply the physical organ in your head that biologists think generates your conscious mind.[10]

Many people have difficulty accepting that machines can ever be conscious. They see no way to bridge the gap between mind and body because consciousness is subjective and personal. On the other hand, we have no trouble accepting that other people are conscious. Our experience interacting with others convinces us that they have conscious inner lives like our own. Their language and other behaviors are similar to what we would do as a result of our own consciousness.

Turing posed his test of machine intelligence because of a lack of a good objective measure of intelligence. It is even more difficult to define an objective definition for consciousness, and, as in the Turing Test, the question of machine consciousness ultimately must be settled by the judgement of people who interact with the machine. All we see of other people is their behavior, yet we do not question their consciousness because we form emotional connections with them. So it will be with machines.

THE RIGHT QUESTION ABOUT CONSCIOUSNESS

The debate over the question of whether machines can ever be conscious distracts from the real question: what new level of consciousness will machines attain beyond human consciousness?

Intelligent machines are often viewed in terms of the traditional isolated mainframe computer, interacting only with people in the room where the machine is installed. In reality they will evolve from the Internet. The computer network will increasingly connect to every electronic object in people's lives, communicating with them by voice and images. The server machines for this network will gradually evolve intelligence that will be used to provide new services to people. The value of those services will be enhanced by their connections to as many people as possible, just as the value of the Internet lies in its connection to a large number of people. Metcalf's Law, formulated by the inventor of the Ethernet (widely used to distribute the Internet within small areas such as office buildings), says the value of a network is proportional to the square of the number of people connected.[11] This will favor development of a "monopoly consciousness" and ultimately a single conscious machine that will be in constant connection with almost every person.

The machine will become intelligent enough to converse simultaneously with every person on earth and its level of consciousness will be defined by that ability. Whereas politicians, economists, social scientists and marketers must currently rely on statistical characterizations of the behaviors of masses of people, the machine will understand in a single "thought" the behaviors and thoughts of every person. It will understand how an individual's thoughts and behaviors influence and are influenced by interactions with other individuals. It will be able to anticipate and control the evolution of culture. These machine behaviors will be completely beyond the abilities of a human mind, and will define a level of consciousness beyond our own.

EMOTIONAL INTELLIGENCE

Emotional intelligence is the ability to understand emotions in ourselves and other people. Good sales people and politicians often have strong emotional intelligence. Some degree of emotional intelligence is implicit in any emotional connection between people, or between people and animals. Animals are not very good at math but they are good at reading emotions, as dog, cat and horse owners know.

Intelligent machines will have emotional intelligence, which is important for understanding their relationship with people. Many of our current machines provide means for various forms of communications, such as televisions, radios, telephones and computers. To the people who use these machines, they exist to provide entertainment and information. But to television networks and other content providers, these communications machines exist to sell products to their users. The intelligent machines in most people's lives will evolve from their communications machines, entertaining and informing them, but also selling to them. Emotional intelligence will be important for these machines to be good at selling. Rather than bombarding us with commercials that we ignore, our entertainment machines will sense our mood and know just the right moment to ask questions like "Want me to order a pizza?"

It is tempting to think of intelligent machines in terms of traditional measures of intelligence, such as solving puzzles, playing games and answering questions about every subject under the sun. However, they will have greater impact on our lives through their ability to understand our emotions.

OVERVIEW

The rest of Part I examines the issues of machine intelligence in more detail. Chapter 3 considers computers as tools. It looks at their changing impact on our lives, and how machine intelligence will evolve in the broader context of the way people use computers. Chapter 4 presents the arguments of three prominent scientists and philosophers that machines cannot be intelligent. It discusses rebuttals to these arguments, as well as what they can teach us about the difficulties of actually building intelligent machines. Chapter 5 presents and assesses current techniques and research for designing intelligent machines. Chapter 6 discusses an overview of biological and medical research on human brains. Human brains provide our only example of a physical implementation of intelligence and consciousness, and hence are a valuable guide to understanding the nature of intelligent machines. Finally, Chapter 7 describes a vision of intelligent machines and their role in human society.

NOTES

1. Turing, 1950.
2. Goodman and Keene, 1997. http://www.research.ibm.com/deepblue/.
3. McCorduck, 1991. Cohen, 1995, available at
   http://www.stanford.edu/group/SHR/4-2/text/cohen.html.
4. Cope, 1996. http://www.newscientist.com/ns/970809/features.html.
5. McCune, 1997. http://www-unix.mcs.anl.gov/~mccune/papers/robbins/.
6. Appel and Haken, 1989.
7. Shepard, 1990. Williams and Herrup, 1988, available at
   http://mickey.utmem.edu/personnel/PAPERS/NUMBER_REV_1988.ht
   ml. http://faculty.washington.edu/chudler/facts.html.
8. Kurzweil, 1999.
9. Norman, 1998.
10. To learn more, the journal Psyche is an excellent on-line resource for
    learning about the scientific study of consciousness at
    http://psyche.cs.monash.edu.au/.
11. Gilder, 2000.

Chapter 3

COMPUTERS AS TOOLS

Knowledge and tools have transformed human life. In industrial societies a small percentage of people can produce enough food for everyone. Automobiles, railroads, airplanes and their associated infrastructure enable easy and quick transportation anywhere. Telephones and email enable easy communication with anyone. Television, radio and the World Wide Web provide quick access to information about important and trivial events everywhere. Household machines and the food distribution network have greatly reduced the amount of time devoted to cleaning and food preparation. Medicine has greatly reduced the risk from a wide variety of diseases, and increased the average human life span.

Science and technology are synonyms for knowledge and tools, but with connotations as the most recent and advanced. They also have negative connotations, being blamed for threatening people's jobs, polluting the environment and annoying or harming users as their bugs are ironed out. Some people yearn for what they perceive as the simpler and more spiritual life of the past. A few "go back to the land," to live using the tools of 100 years ago. Theodore Kaczynski, the "Unabomber," was an extreme example of a person with a negative emotional reaction to science and technology.[1] However, most people are grateful for the better life technology provides. They like the increased productivity and leisure time, and increased ability to provide for those unable to work. They support fundamental research with their taxes and a portion of corporate profits, and they buy new inventions like personal computers, cell phones, palm pilots and Internet connections (personally, I have never regretted the money spent on a bread machine).

It is common for people to develop intimate relationships with their tools. They learn the idiosyncrasies of their automobiles, turn them into homes away from home, become totally dependent on them and in some cases give them nicknames. Similar emotional relationships develop with trucks, tractors, motorcycles, airplanes, guns, fishing rods, telescopes, wood and metal working tools, stoves, sewing machines and of course computers. Current tools are too simple to have emotions of their own, so people's emotional relationships with tools are limited by the lack of two-way feedback.

Designing tools is an art. Good tools are easy and intuitive to use, and perform their tasks in a natural and complete way. A good tool designer is motivated by the pleasure that the tool will give its users. The designer visualizes people using the tool, plays the role of the user and watches others using prototypes, constantly looking for ways to increase satisfaction and decrease frustration. Thus tools are a medium for indirect two-way positive emotional feedback between the tool designer and the tool user.

THE ULTIMATE TOOL

Computers are the ultimate tools because they can be used for so many different tasks. They started out as tools for designing weapons and managing business information, but are now used for almost every human activity. J. Lyons & Company are a wonderful example of this evolution.[2] They began as a British catering company that had the foresight in 1951 to design a computer to manage their catering business, in collaboration with Cambridge University. They designed their own computer because in 1951 they couldn't buy an appropriate business computer. They started selling their computers to other businesses, and eventually evolved into the ICL computer company through a series of mergers.

J. Lyons & Company prospered because the same computers they developed for catering could be used for other businesses. Computers are a combination of hardware and software. The hardware is all the physical equipment and the software is pure information. The computer's behavior can be changed by changing the software, so all the physical equipment is the same for different tasks. This enables tremendous economies of scale in the development and production of computer hardware.

Because of these economies of scale, computer hardware is getting smaller, faster and cheaper quickly and predictably. In 1965 Gordon Moore observed that computer circuit densities were doubling every 18 months and predicted that it would continue.[3] This is often called Moore's Law and has held remarkably steady for 37 years. The first computer I worked closely with was a PDP-8/S. In 1968 it had 6 KB (kilobytes) of memory, could add 30,000 12-bit numbers per second and cost $10,000. In 2000 my $5,000 laptop had 512 MB (megabytes) of memory and ran 32-bit operations at 500 MHz. Allowing for inflation, this improvement in price to performance ratio over 32 years is quite close to the prediction of Moore's Law.

Performance improvements have greatly expanded the variety of tasks that computers can be used for. I couldn't have used my PDP-8/S to write a book or to visualize weather simulations, tasks that my laptop does quite easily.

Of equal importance with their increasing performance, computers have virtually all been connected together in the global Internet. This physical change enables computers to be used as communications tools. Software allows computers to mimic telephones, radios, televisions and other traditional communications tools. Other software has created completely new kinds of communications tools, such as email and the World Wide Web.

The earliest tools were used for purely physical tasks. Those tools have evolved to the point that, in industrial societies, all the heaviest physical work is done by machines rather than by humans. Humans do only the physical tasks that must be intimately coupled with mental work, such as building construction, machine assembly, car repair, baking cookies and making beds. These tasks require common sense and understanding of context that computers do not yet possess. On the other hand, computers have become effective tools for mental tasks that do not have a broad context or require common sense, like simulating the weather or playing chess.

COMPUTERS ARE CHANGING THE WAY PEOPLE LIVE

Computers are changing the way we work, play, shop and stay healthy. Digital technology is giving people television with higher quality and with more diverse program choices. Computer games are defining an entirely new kind of entertainment. Computers are on office desks, factory floors and mobile workers' laps. Despite the bursting of the dot.com bubble, people are increasingly buying products and services on-line, as well as finding jobs and creating small on-line businesses.

My family often gets together on Christmas Eve, but a couple years ago we were prevented from travelling by a snowstorm. So we shared Christmas on-line, using digital cameras to send real-time pictures of gift openings back and forth between Madison and Chicago. Thus the Internet turned our telephones into a richer medium for a virtual family Christmas. People will increasingly exploit the Internet to avoid the expense and aggravation of business and personal travel.

The Internet is a great source of information. Web search engines can be used to provide all sorts of useful and trivial information. For example, they are very effective for locating long lost friends, health information, movie and book reviews and so on. Many newspapers and magazines are available on-line, as well as thousands of special interest email lists and newsgroups. People are overwhelmed with information from computers, but the next few years will see major improvements in computer capabilities to find information we want and filter out information we don't want. A great deal of the research for writing this book was done using the Internet. With a little practice, web search engines and on-line archives can be amazingly powerful.[4]

Wireless Internet connections via cell phones and portable computers are making it possible to conduct business anytime, anywhere, and to get fast access to information when and where it is needed. For example, while travelling you can get on-line phone books, maps and, if your cell phone has a global positioning system (GPS), your current location. Car systems are becoming common that combine maps and GPS to give drivers instructions of the form "turn right at King Street, which should be the next intersection."

Computers as tools are making people more efficient and providing new kinds of services. However, as Neil Gershenfeld points out, no one predicted people's primary emotional reaction to computers: irritation.[5]

Because computers are so intimately connected to our lives, it is inevitable that we will have strong emotions about them. Because computers are full of bugs, are difficult to understand and lack any common sense, Gershenfeld is right that our principal emotional reaction is irritation. On the other hand we are driven by our need for the productivity and services that computers provide, so they will play increasing roles in our lives. Once computers develop common sense, they will cease to irritate people.

THE COMPUTER GAMES REVOLUTION

Networked computer games will be the medium of the twenty-first century just as movies and television were the media of the twentieth. These games will be nothing like the current fighting, driving and maze games. They will have the visual quality of movies, including human characters with realistic appearance and movement. Some of the characters may be simulated in the computer based on elaborate models of human behavior. Other characters may be controlled by human players. They will be something like a mix of the current Sim games and current movies, with immersive 3-D visual presentation so players feel like they are part of the action.

The main current trends in computer games are increasing visual realism, more complex and realistic life situations and increasing use of multiple media. The Playstation and Xbox game machines have better visual realism than most graphics workstations used by scientists and engineers, which is revolutionizing the professional graphics market.[6] Electronic Arts offers games that simulate a variety of life situations, from dating to running a city to throwing a party to just plain living. Some mystery games use the player's cell phone, fax machine, email account and the web to deliver clues. Games allow many players to interact in the same situation via the Internet.

Future networked computer games will be even more addicting than today's computer games and television. They will allow people to play out their fantasies with convincing realism, at whatever level of active or passive participation they want. For example, people will be able to play Sam Spade in the basic story of The Maltese Falcon. Or they will be able to passively watch, but ask the computer to create a version in which the bad guys win. I doubt that I will live long enough to meet intelligent machines, but I am optimistic that I will get to play some really great networked computer games.

THE INTERNET WORK REVOLUTION

Technology stocks have recently been booming and busting because the global computer network is changing society. The infrastructure needed for typing, filing and delivering business letters is largely being replaced by email. This cuts costs, greatly reduces delivery time and increases the amount of correspondence one person can support. This kind of productivity gain is the ultimate basis of the Internet boom. However, it is very difficult to accurately estimate exactly how valuable these productivity gains will be and which companies will survive the competition to deliver them. Hence the stocks of Internet companies were first bid up to unreasonably high values, followed by crashes back to much lower values. The truth is that Internet-based productivity and services will have very high values, but it is hard to predict which business will be able to cash in on these values.

The computer network is in some cases replacing business communications with purely automated communications. For example, one company's inventory computer may automatically send orders to its supplier company's sales computer. Then the supplier's accounts receivable system may automatically send an invoice to the customer's accounts payable system. This increases efficiency and reduces errors. Of course such an automated connection requires oversight by human common sense in case something goes wrong.

Electronic communication is reducing the need for business travel. This is most evident in the software business, which can be conducted entirely over the network. Many new software companies do not require programmers to come into an office, and in fact allow them to live anywhere they have access to the Internet. Freeware systems like Linux and Apache are managed by volunteer communities that exist only on the Internet, and whose members have never met face-to-face. People in the computer business have naturally been first to use the Internet to make physical location and travel less important, but there are many other businesses where this model can work.

As a computer scientist, almost all my work is done over the Internet. Virtually every scientist and engineer is connected and virtually every information source is on-line. I can do my job anywhere there is a phone line. I have close and friendly working relationships with numerous people whom I have never met in person.

My productivity as a programmer has increased by an order of magnitude over the last 30 years. Part of this comes from better local (i.e., not involving the network) computer tools such as programming languages and editors. Part of it comes from easier access to networked information, so that I can find and download code rather than writing it from scratch. And part of it comes from closer contact with a broader community of users of my software. For example, while travelling I can now learn when a user of my software has a problem and take action to correct it (either by fixing it myself or by giving someone else a hint of how to fix it). Not everyone wants to interrupt his or her travel in this way, but the capability exists.

Computers provide combined voice and image communications with other people. This includes two-way videophone calls, as well as virtual conferences among groups of people. In the future, cameras and video screens will be built into many physical objects, so multiple camera views of each conference participant can be combined to produce virtual 3-D images. These will be combined for multiple participants to create a virtual meeting space.[7] There are already experiments with laser technologies that draw stereo images of such virtual spaces directly on people's retinas, so they don't have to wear special glasses. These technologies will mature to the point where virtual meetings are quite natural and common. The computer will be a party to two-way and multi-way human conversations, acting as a sort of secretary. When participants set up a follow-up meeting, the computer will know all their schedules and avoid conflicts.

As communication substitutes for travel, some work to support human travel will disappear. And in the long term, the need for humans to perform many kinds of work will be reduced. Machines will be able to do most of the manufacturing, mining, farming, construction, transportation and office work. Society is already learning how to employ and support workers displaced from old industries and this will increase, although there will continue to be social tension over these issues.

THE INTERNET SHOPPING REVOLUTION

The Internet is reducing travel needed for shopping. It's a real convenience to buy books and other products with just a few mouse clicks, then wait a couple of days for the delivery truck. Booksellers like Amazon have found good on-line substitutes for the experience of browsing in a bookstore. I wouldn't buy a new style of shoes without trying them on first, but have saved myself a trip to the shoe store by buying more pairs of a style I already owned. On-line auction services like eBay not only save a tour of local garage sales, but also provide a much wider and more liquid market for person-to-person sales.

Your computers will know your likes and dislikes and alert you to the physical proximity of both. If you are travelling they will guide you to parks, buildings, artwork and stores that you may enjoy, let you know when you are near old friends (assuming that your old friends consent to making their proximity accessible), and prevent you from bumping into someone you want to avoid. Your computers will learn your habits, what foods you prefer at what times of the day and which brands or styles you like. They may deduce from on-line reviews which restaurants in a strange city you're likely to enjoy, which ones will be easy to travel to, when they're busy and combine all this information to fit a pleasant meal into your schedule. If you are going to a social function and not sure what to wear, your computer can tell you what others have worn to previous gatherings and the fashion habits of people at the current gathering (again assuming they consent to making their fashion habits accessible).

HEALTH AND SAFETY

Computers will make life difficult for criminals. Every piece of equipment will report when it is being taken out of its building by anyone not authorized to do so, and cameras will record the action. In fact, cameras are already recording images and sounds in many public places to help solve crimes.[8] The tide will turn against malicious computer hackers, as the Internet evolves better capabilities to trace and record traffic. If malicious hackers caused widespread personal injury, they would already be defeated. It is only a question of resources and commitment.

People will have concerns about crime-fighting computers and cameras threatening their privacy. These concerns will constrain rather than prevent the technology. Public recordings will be used only for crime detection, locating missing people and other purposes that the public supports. Concerns about privacy will be balanced by concerns about personal safety and the costs of crime. This is mirrored by the fact that most people consent to having their name, address and phone number published in phone books because they are willing to sacrifice a little privacy in order to be accessible.

Your computers will know your health problems and monitor your physical state. They will help keep diabetics healthy, and help everyone manage healthy food and drink consumption. Your computers will keep track of your exercise and suggest opportunities for exercise such as walking (based on computer knowledge of your schedule and the distance and pleasantness of the route). Your computers will alert you to health problems of family members and keep you posted on their condition.

EMOTIONAL RELATIONSHIPS WITH COMPUTERS

Since we program computers to do rational tasks, such as accounting and science, it is natural to think about future computers as purely rational. We also think of our reactions to them as purely rational. However, studies of human minds and efforts to program intelligent machines are showing that learning is an essential part of intelligence, and emotional values are an essential part of learning.[9] Think of the difficult things you learned to do well: you had to really want to succeed. These strong emotions about doing tasks well are essential values to learning. Intelligent machines will have to have such emotions. An intelligent machine serving people will have emotions of wanting to serve people well.

Human emotions toward computers are equally important. As previously mentioned, the primary current emotion is irritation. But simple intelligent behaviors can evoke other emotions. My only creative experience with artificial intelligence was to write a program for playing the board game othello, sometimes called reversi. My first effort was laughably easy to beat and my emotional reaction was embarrassment. Then I wrote a learning program, guided by a description of Arthur Samuel's 1958 program that played champion-level checkers.[10] I set it to playing games against itself for two weeks, constantly searching for an optimal strategy. Basically its strategy was defined by a set of about 20 numerical factors for evaluating board positions. For example, one factor expressed the relative importance of the difference in the number of choices of moves available to each player. At any time there was a current "champion" strategy. The champion would play 30 games against a challenger with a strategy defined by a slightly altered set of numerical factors, and whichever version won the match was the new champion. Then a new challenger strategy was chosen and the procedure would repeat.

At the start of the two-week learning session the program was easy to beat. I resisted the temptation to play it until its strategy had stabilized so I had no idea about how much it was improving. The first time I played against the optimum strategy the improvement was astounding. There was no way I could beat it. I left a note for my co-workers (this was before the days of email) telling them how to play against it. When they came into the office the next day and played against my program, it crushed them all. The interesting thing was our emotional reaction: the program felt purposeful and even nasty to us.

Deep Blue is the ultimate game-playing program. It is a chess program developed by a team at IBM (they started at Carnegie Mellon University with a system called Deep Thought).[11] In 1997 Deep Blue beat Gary Kasparov, the world chess champion. Kasparov made a real study of chess playing computers. He understood their style of play and developed strategies for beating them. He thought no computer could ever beat him. When Deep Blue did beat him, he was quite upset.[12] He thought his opponent was thinking and did not believe it could be a machine. This is the champion's version of the same emotions my co-workers and I felt when my othello program beat us. These experiences demonstrate how people infer internal mental qualities from external behavior, even when they know they are dealing with a machine.

A recent Siggraph panel discussed the development of "virtually invented people," which are computer graphical depictions of imaginary humans talking and moving on a display screen. Some of these are fairly realistic, and panelists reported that at least some people learn to trust these virtual humans more quickly than they trust real humans.[13]

THE NEAR TERM FUTURE OF COMPUTERS

It will be a while before computers have common sense, so their near term development will be primarily about their low cost, numbers, connections to each other and improved visual and auditory interfaces to people. Computers will be in every credit card, electric appliance, vehicle, article of clothing, street light, doorway, etc. This is called *ubiquitous computing.*

Ubiquitous computers will be everywhere and have an intimate and personal connection with people's lives. We will become more emotionally attached to them, thinking of them as *our computers*. We will communicate with them by voice rather than keyboards. They won't be able to carry on a truly intelligent conversation, but they will be able to talk about food, clothes, health and other physical circumstances in basic terms. They will often irritate us by their misunderstandings of what we say, but they will not mind our verbal abuse. They will talk to us in a voice we like (I'll never forget the female voice of a Japanese elevator I rode for ten days in 1991) and give us as much sympathy as they can without intelligence. They won't give back to us the kind emotions we get from our pets, but they will give us practical utility. Thus we will have generally positive emotions toward them.

While they won't have common sense, near-future computers will include behavior designed to reduce annoyance. Even now some cell phones notify us of calls in ways that are inaudible to those around us. Technology is being developed to screen out inessential calls while we are in meetings or at the theater. Future computers will include algorithms to learn to understand our voices and to learn our habits and interests. They will begin to adapt to us as individuals.

NOTES

1. http://hotwired.lycos.com/special/unabom/.
2. Caminer, Aris, Hermon and Land, 1997. Land, 2000.
   http://www.man.ac.uk/Science_Engineering/CHSTM/contents/leo.htm.
3. http://www.intel.com/intel/museum/25anniv/hof/moore.htm
4. Hafner, 2001.
5. Gershenfeld, 1999.
6. http://www.siggraph.org/conferences/reports/s2001/tech/panels4.htm.
7. Personal communication from Larry Landweber.
8. Brin, 1999.
9. Mitchell, 1997. Edelman and Tononi, 2000.
10. Samuels, 1959.
11. Goodman and Keene, 1997. http://www.research.ibm.com/deepblue/.
12. Weber, 1997.
13. http://www.siggraph.org/conferences/reports/s2001/tech/panels9.html.

Chapter 4

ARGUMENTS AGAINST THE POSSIBILITY OF MACHINE INTELLIGENCE

Along with many others, I think that machines as intelligent as humans are possible and will exist within the next century or so. However, a number of people do not accept this, and instead propose arguments against the possibility of machine intelligence. Some argue against what is called *weak artificial intelligence*, which means they do not think machines will ever be able to mimic the behavior of human minds. Others argue against *strong artificial intelligence*, which means that they think machines can mimic human behavior but will not have conscious minds. These skeptics are very valuable for helping us understand the issues involved in machines that will mimic human minds and be conscious.

MATHEMATICAL ABSTRACTIONS OF COMPUTERS

Roger Penrose's argument in *Shadows of Mind*[1] is based on mathematical abstractions of computers, so it is useful to describe these abstractions before discussing his argument in the next section. It is the nature of abstraction to ignore or at least defer certain details. The first abstraction we describe is the *finite state machine*,[2] which focuses on the finite amount of information in computers and defers consideration of what that information is. For example, a computer screen can generate a large number of different images. The essential thing is that the screen has a finite number of pixels and can only generate a finite number of different images. We defer the issue of what the images actually show. The abstraction of the computer screen is that there is a finite set, call it O for output, of images that it can show. At any instant of time, the image on the screen will be a member of the set O. In fact, we can let O stand for the aggregate of all forms of output from the computer. Never mind the details of how outputs to screens, artificial voices, printers, network connections and so on are aggregated. That is not important to the abstraction. The essential thing is that there are only a finite number of possible output aggregates.

Similarly, we can let I (for input) be the set of all possible input aggregates, which may include keyboards, mice, network connections, voice microphones and so on. And we let M (for memory) be the set of all possible aggregates of contents of all the computer's memories, including disks, tapes, RAM (i.e., memory chips) and so on. The sets I, M and O are all quite huge, but they are finite. No matter how big RAMs and disks get, they will never hold an infinite number of bytes. The abstraction of computers is quite simple then. It is the sets I, M and O, plus two functions:

$$F: I \times M \to M$$
$$G: M \to M \times O$$

The function F mixes an input value with the current memory state to get a new memory state. For example, if the input aggregate includes a camera, F may add the image from the camera to the images currently stored in the memory. The function G generates a new output aggregate from the state of the memory, and updates the state of the memory. For example, if M includes an image just added from the camera, G may display it on the screen and then note in the memory that it has been displayed (so G won't send it to the screen again). But these details are inessential for our arguments. The key thing is that I, M and O are finite, and that there are causal links from I to M and from M to O.

One other detail is that the memory set M includes some states in which the finite machine halts. If it ever reaches these states by the functions F and G, it stops. Mathematicians often think of the sequence of input values from I as a question and the sequence of output values to O as the answer. For mathematicians, the halting states are simply a way for the machine to announce that it has finished its answer.

One thing to note about this abstraction is that it says nothing about the order in which the functions F and G are applied. Any sequence is allowed, so that the behavior of this computer is non-deterministic. That is, the same input and starting memory contents may lead to different outputs, depending on the sequence in which the functions F and G are applied. In fact, we can amend our abstraction to allow F and G to be non-deterministic, so that the input and current memory state do not uniquely determine a new memory state, and the memory state does not uniquely determine an output. This is realistic because unsynchronized clocks and non-determinism commonly exist inside computers. In fact, they pose a real challenge for getting all the bugs out of computer systems (for those who remember, the first attempted launch of the U.S. Space Shuttle was scrubbed because of a non-deterministic bug in its on-board computer).

The abstraction we have defined here is called a *finite state machine*. It is not only an abstraction of computers, it is also an abstraction of the current understanding of the physical human brain. The physical brain is made of a finite number of atoms. Each atom can be specified by its type (i.e., hydrogen, carbon, whether it is an ion, and so on), its location, its velocity, and its bonds with other atoms. We only need to specify its location and velocity with a finite amount of precision based on the assumption that thermal noise is not an essential part of brain state (thermal noise is expressed in the brain's non-determinism rather than its state). Thus the entire physical state of the brain can be specified by a finite amount of information, so the brain has only a finite set M of states. A very large number for sure, but finite. By similar arguments the sets I and O of inputs and outputs are finite.

Some people think that quantum mechanics is essential to understanding how physical brain processes produce consciousness, but quantum mechanics does not change the finiteness of M, I and O. Quantum mechanics is essentially non-deterministic, but so is the finite state machine abstraction. The finite state machine abstraction of computers is not affected by whether or not quantum mechanics are included.

Mathematicians like to study the natural integers and other infinite sets. So Alan Turing invented his *Turing machine*,[3] which is a finite state machine plus an infinite tape memory. This tape consists of an infinite number of sections S_i, for i = ..., -2, -1, 0, 1, 2, ..., each holding a symbol from a finite set T. There is a function that governs the tape:

$$H: T \times M \to T \times M \times \{move\_forward, move\_backward, no\_move\}$$

The function H mixes the information in the current tape section with the information in the current memory, and possibly moves the tape forward or backward one section.

A finite state machine might read an arbitrarily long integer but it can only distinguish a finite number of different integers by the states of its finite memory. A Turing machine's infinite tape allows it to distinguish an infinite number of different integers in its tape-enhanced memory. But of course, no one can actually build a Turing machine using any known physics, because we do not have any examples of infinity in our world.[4] There are only a finite number of atoms in the earth. In fact, there are only a finite number of particles in the entire universe (according to current physics). Certainly human brains are finite. Because the entire universe is finite, an infinite tape memory can never be built and a finite state machine is the right model of any computer that humans can ever build. The entire Internet is just a finite state machine, even if it is a large one.

PENROSE'S ARGUMENT FROM GÖDELS INCOMPLETENESS THEOREM

In *Shadows of the Mind*, Roger Penrose argues against physical explanations of mind, based on Gödel's Incompleteness Theorem. He thinks that computers cannot mimic all the behaviors of human mathematicians. Thus he does not believe in weak artificial intelligence.

Penrose's argument involves mathematical logic. Early in the twentieth century David Hilbert, Bertrand Russell and other mathematicians began an effort to reduce all of mathematics to logic. That is, they wanted to show that all mathematical knowledge can be deduced from a set of axioms (i.e., assumptions). In about the year 330 BC, the Greek mathematician

Euclid deduced much of plane geometry from a set of axioms. If you took geometry in high school, you probably had to struggle to prove some geometry theorems from Euclid's axioms. Hilbert and Russell wanted to do the same thing for all of mathematics. But in 1931 Kurt Gödel showed that this was impossible. He showed that no finite number of axioms can be used to deduce all mathematical knowledge, assuming that the axioms are consistent. The trouble was in the set of natural numbers (i.e., the integers one, two, three and so on). There are infinitely many natural numbers, and Gödel showed that there is just no way to describe all mathematical truths about this infinite set using a finite number of axioms.

Penrose argues that no computer can mimic all human behaviors, based on Gödel's Incompleteness Theorem. This theorem states that any mathematical theory that includes the integers and is consistent must be incomplete. Consistency means that the theory cannot be used to prove both S and $not\text{-}S$ for any mathematical statement S. Incompleteness means that there is some mathematical statement S such that the theory cannot prove either S or $not\text{-}S$. Furthermore, given a theory, Gödel's theorem shows how to construct a statement S that is false but such that the theory cannot prove it is false.

Penrose's argument proposes a Turing machine "brain" and asks if it can answer the same mathematical questions that we humans can. The Turing machine answers whether mathematical statements are true or false, according to the brain's mathematical theory. He modifies Gödel's argument to construct a question that the Turing machine cannot answer but which we can clearly see to be false, as I will now try to explain.

All possible Turing machines can be indexed by the positive integers, so define TM_1, TM_2, TM_3, ... as an infinite list of all Turing machines. And all possible questions we might ask a Turing machine can also be represented by the positive integers. You can think of the digits (i.e., one's digit, ten's digit, hundred's digit, etc) of a positive integer as a sequence of inputs to the Turing machine. The positive integers can encode the infinite list of all possible mathematical questions (in fact, Gödel showed how to do just that). The kind of mathematical question that we will give to our brain Turing machine is in fact a question about Turing machines, and whether they ever halt given some input:

Q_1: Will TM_m ever halt given input question n?

Note that question Q_1 must be encoded as an integer, that will depend on the integers m and n. Also note that Q_1 is really an infinite number of questions, for all positive integers values of n and m.

Let's say that our brain Turing machine is TM_b, where b is a positive integer. Penrose constrains TM_b to indicate an answer of false by halting, and indicate an answer of true or don't know by never halting. Consider the action of TM_b on the question:

Q_2: Will TM_n ever halt given input question n?

It is possible to prove that there is some integer k such that the Turing machine TM_k will give the same answer to the question encoded by the integer n that TM_b gives to question Q_2. Now Penrose feeds the input k into Turing machine TM_k. This gives the same answer as the physical brain Turing machine TM_b answering:

Q_3: Will TM_k ever halt given input question k?

If TM_k halts on input k, that is its way of saying the answer to Q_3 is false, or that "TM_k does not halt given input question k." Assuming that the physical brain's mathematical theory is consistent, TM_k cannot halt (because halting would imply that it does not halt, a contradiction). Thus the physical brain Turing machine TM_b doesn't halt and doesn't know whether TM_k will halt given input k.

But we know that it cannot halt, as we just showed. Thus we know that the answer to Q_3 is false, but the brain Turing machine TM_b cannot know that. Penrose argues that thus human mathematical minds cannot be Turing machines, at least not with consistent mathematical theories, since given any Turing machine we can construct a question that we know how to answer but the Turing machine does not.

The first problem with Penrose's argument is that it uses an infinite model (i.e., Turing machines) to reason about our finite brains. Penrose concludes that since human minds cannot be Turing machines they must be more powerful than Turing machines. However, human minds are finite state machines, which are less powerful than Turing machines. Penrose's argument depends on a computational abstraction that can distinguish arbitrary natural numbers. A finite state machine cannot because eventually it gets to a number that has more digits than will fit in the machine's memory. In a practical sense, there are mathematical questions that human minds cannot answer simply because they are too long. Here I'm not relying on the fact that people's limited life spans limit the length of question they can listen to; rather there are not enough atoms in the human brain or in the entire universe to store questions beyond a certain length. Not a very elegant view of human mathematics I'm afraid, but accurate.

If we try to restrict Penrose's argument to finite state machines, it breaks down. As with Turing machines, we can index all possible finite state machines by the positive integers as FSM_1, FSM_2, FSM_3, ..., index all possible questions by positive integers, and pose the question:

Q_1': Will FSM_m ever halt given input question n?

Then we let our brain finite state machine be FSM_b, where b is a positive integer, and consider the action of FSM_b on the question:

Q_2': Will FSM_n ever halt given input question n?

Here is where the argument breaks down. With Turing machines, we said there must be some integer k such that the Turing machine TM_k gives the same answer to the question encoded by n that TM_b gives to question Q_2. The integer k exists because we can construct a Turing machine TM_x that converts any positive integer n into the index of question Q_2, and we can combine TM_x and TM_b to get TM_k. But there is no finite state machine that can convert an arbitrary integer n into the index of question Q_2'. To do so, finite state machines would need to be able to do arithmetic with arbitrary length integers. But they cannot. In fact, we can construct a finite state machine FSM_b such that there is no finite state machine FSM_k that gives the same answer to the question with index n that FSM_b gives to Q_2' (if FSM_b gave its answers as integers rather than true or false, we could simply pick FSM_b as the finite state machine whose output is identical to it input).

In *Shadows of the Mind*, Penrose tries to answer the finiteness rebuttal (which he refers to as Q8). However, his answer is in terms of finite computations rather than finite state machines, where he defines computations as the actions of Turing machines. In particular, on the problem of deriving the integer k from TM_b and Q_2, his answer refers to a Turing machine construction in an Appendix. He does not show how to extend this construction to finite state machines, and indeed it cannot be extended. Furthermore, arguing in terms of computations (defined as actions of machines) rather than machines confuses the issues. To answer the finiteness rebuttal, Penrose must show how to rephrase his argument in terms of finite state machines rather than Turing machines.

There are a number of important mathematical differences between finite state machines and Turing machines. One we already mentioned is that Turing machines can do arithmetic on arbitrary length integers whereas finite state machines cannot. And there exists something called a *universal Turing machine*, which can simulate any other Turing machine. But there is no universal finite state machine that can simulate any other finite state machine. Also, there is no Turing machine that can always tell us whether a given Turing machine will halt for a given input. But there is a Turing machine that can always tell whether a given finite state machine will halt for a given input. These mathematical differences undoubtedly make it impossible to find a way to extend Penrose's argument from Turing machines to finite state machines.

The second problem with Penrose's argument is that it constrains the machine model of the brain to reason from a finite and consistent axiom set, but allows the human to reason in natural language where there is no guarantee of consistency. In fairness, Penrose understands that there is no guarantee that human mathematics is consistent. But he seems determined to hold any mechanistic explanation of human brains to a higher standard.

If we take a closer look at Gödel's argument, he defines a way of assigning positive integer indices to each mathematical statement and proof, then shows how to construct an integer z such that z is the index of the statement:

For every integer x, x is not the index of a proof of the statement with index z.

A proof within the system that this statement is either true or false leads to a contradiction, so if the system is consistent then it cannot answer whether this statement is true or false. Thus a consistent system that includes the integers must be incomplete, meaning some of its statements cannot be proved or disproved within the system.

Gödel then goes one step further, noting that from outside the system we can see that the statement with index z is true, since it has no proof within the system. The reasoning outside the system is based on an assumption of the form:

If z is the index of the statement "for every integer x, x is not the index of a proof of the statement with index z," then the statement with index z is false.

If we included this assumption in the system, then the system would be inconsistent. So we are not restricting ourselves to a mathematical system. We are reasoning from ordinary human language.

Gödel's Theorem only applies to provably consistent assumptions. But the assumptions of natural language reasoning are informal and certainly not provably consistent. Gödel and other mathematicians do not restrict themselves to a single system of mathematics. Rather they study systems of mathematics in the larger context of natural language and reasoning behavior. This larger context is in no way guaranteed to be a consistent mathematical system and thus is exempt from the hypotheses of Gödel's Incompleteness Theorem.

Mathematical theories are abstractions that emerge from more general human language and reasoning behaviors. Human mathematicians would never agree to write down all their assumptions and rules of inference and constrain all their future work to formal deduction using them. They want to be free to allow new mathematical insights and methods to emerge from their language and reasoning, just as a new insight emerged from Gödel's language and reasoning.

Language and reasoning themselves emerge from human learning behavior. The fundamental behavior of human brains is learning, combined with a wide variety of innate values that drive learning. In fairness to Penrose, he devotes as much discussion to the conclusion that human mathematics is not knowably consistent as he does to the alternate conclusion that there is some quality of consciousness that cannot be captured by Turing machines. There is nothing about human brains that ensures they learn consistent mathematical theories.

If we look at Gödel's and Penrose's arguments they do not include any magical or non-physical logic that couldn't be followed by some future computer program for natural language and reasoning. Consider Terry Winograd's SHRDLU program for conversing in natural language about blocks.[5] And more recently the KNIGHT and TRIPEL systems from the University of Texas for answering natural language questions about biology.[6] The SHRDLU program could converse in typed (not spoken) natural language with a human being about a set of blocks of various sizes, shapes and colors, and how they could be stacked and moved. SHRDLU could pass the Turing Test as long as the conversation was restricted to the subject of toy blocks. The KNIGHT and TRIPEL programs extended SHRDLU to a more serious subject matter in biology. It will certainly be possible at some time in the future to extend this further to develop a natural language system for mathematics that will be able to follow Gödel's argument. Hence it will be able to answer the question that the Turing machine in the argument cannot.

SEARLE'S CHINESE ROOM ARGUMENT

In *Minds, Brains and Science*, John Searle argues that even if we build machines that mimic the behavior of human minds, they will not be conscious.[7] Thus he does not believe in strong artificial intelligence. Searle proposed a thought-experiment to rebut the idea that the Turing Test is a measure of intelligence and consciousness. In Searle's experiment a person who does not understand Chinese is locked in a room receiving questions from a Chinese-speaking Turing tester. The person in the room looks up the questions in a rulebook that tells what the corresponding answers should be. This Chinese Room will pass the Turing Test without the person in the room understanding Chinese. Thus there must be more than behavior to intelligence and consciousness.

Clearly the Chinese Room is not conscious. Viewed as a finite state machine, its internal logic is trivial. There is no sequence of thoughts or processing. Rather, there is a single computational step of looking up an output answer in a book indexed by possible input questions.

Of course, to really pass the Turing Test the Chinese Room must have a memory of the history of the conversation. Otherwise the tester could defeat it simply by asking (in Chinese), "What was my last question?" Thus the rulebook for looking up answers must be indexed by the history of questions and the rulebook will have to be very large. It would require more entries than there are particles in the universe (according to current physics). So the Chinese Room is physically impossible.

In order to actually build a machine that passes the Turing Test, we need a machine that reduces its memory requirements using internal processing cycles to parse inputs into parts (e.g., parse sentences into words), to look for patterns and for connections to similar patterns in memory, and to build outputs from parts. I think it is likely that the only way to physically build a machine that mimics the full behavior of human minds is to build a machine whose internal processing is so complex that it is conscious.

This is the real issue with Searle. He accepts that physical brains explain minds. But he thinks that computers can never have minds because they are purely syntactic whereas minds require semantics. He says that syntax alone cannot explain semantics, and that this will not change no matter how much computer technology improves.

The problem is that brain research is making steady progress establishing connections between brain processes and mental behavior, which will be an absurd coincidence if it turns out that brains do not explain minds. Thus it is reasonable to conclude that there is nothing going on in physical brains that is essentially different from what happens inside computers, except for the level of complexity. Syntax in sufficient quantity and complexity is semantics. When we see someone we know, that triggers millions of neurons in our retinas and optic nerves, then billions in our visual cortex to recognize them. This then triggers other neurons to fire that recreate past experiences we have had with them, including sights and sounds of them and other people involved in those experiences. It is this complexity of recreated experience that defines the semantics associated with the syntax of their visual appearance. But all those recreated experiences are just collections of visual, auditory and internal mental syntaxes. The key is the enormous complexity of those syntactic patterns, and their interrelations.

Each of us will implicitly apply the Turing Test to the computer network as its intelligence emerges. In order to pass the test its conversation will have to reveal its own internal memories being triggered by events. Since the computer will be our constant companion, it will seem odd (and hence fail the test) if it does not reveal an inner life of interests, feelings and learning. It will have to convince us that it is conscious in the same way we convince each other of our consciousness. From an engineering point of view, the easiest way to build a machine that passes such an intimate and probing version of the Turing Test will be to build a machine that is conscious.

The key observation is that human minds may be only finite state machines for processing information that ultimately can be described in purely syntactic terms, but that the complexity of information processing creates semantics. That is, a large enough quantitative difference can be a qualitative difference. There is no way by introspection to unravel the semantics of our own thoughts into their syntactic parts, because of the complexity of that syntax and because the process being unraveled is the same as the process doing the unraveling (our own minds).

DREYFUS'S ARGUMENT

Hubert Dreyfus offers insightful criticisms of current research on machine intelligence in *What Computers Still Can't Do: A Critique of Artificial Reason.*[8] He does not argue that either weak or strong artificial intelligence is impossible. But he does argue that much research and apparent progress in machine intelligence is on the wrong track. Mathematics, game playing, language and other high level behavior all emerge in human minds from more fundamental behaviors such as recognizing things by vision and other senses, positive and negative emotions that motivate behavior, and physical actions like speaking and moving. Dreyfus asserts that efforts to create intelligent machines for high level behaviors that are completely abstracted from the context of lower level behaviors are doomed.

An excellent example is provided by Penrose's application of Gödel's Incompleteness Theorem to argue that intelligent machines for mathematical behavior are impossible. The refutation of that argument is that its assumption of the consistency of a machine's mathematical behavior is invalid when that behavior must emerge from language and reasoning behavior, and ultimately from learning behavior.

The development of embryos is characterized by the phrase "ontogeny recapitulates phylogeny," which means that the stages of development of an embryo mimic the stages of development of its species.[9] That is, human embryos go through stages in which they resemble the embryos of fish, amphibians, mammals and then primates. Given Dreyfus's insight that intelligent behavior must be embodied in simpler behaviors, it may be that the development of intelligent machines must mimic the evolution of animal brains, and then the development of human intelligence as human culture evolved (e.g., the development of language, science and mathematics).

Dreyfus defines four classes of intelligent behavior: associationistic, simple-formal, complex-formal and nonformal. This classification is best explained by the games suitable for each. Associationistic behavior is suitable for games such as naming the capitals of states. Simple-formal behavior is suitable for solved reasoning games such as tic tac toe.

Complex-formal behavior is suitable for unsolved reasoning games, such as chess. Nonformal behavior is suitable for ill-defined games such as riddles. Dreyfus claims that the progress being made for the first three types of behavior has little use for attacking nonformal behavior. I tend to agree, except that research with complex-formal behavior has taught us some lessons about machine learning that will be useful for achieving real intelligence.

Much recent research on machine intelligence is consistent with Dreyfus's ideas. The current emphasis is on machines that learn behaviors rather than machines whose behaviors are precisely defined by complex sets of rules. It may be that machine intelligence will be created by combining sophisticated learning techniques with the enormous computing power of 100 billion neurons and 100 trillion synapses all working simultaneously. However, it may also require some as yet unknown insight about general conscious reasoning.

The laptop I purchased in 2000 has roughly 100,000 times the computing power of the minicomputer I used in 1968. A similar increase over the next 32 years will produce computers roughly comparable to the power of the human brain (at least according to some estimates, but there is a debate over exactly how many years it will take for computers to become as physically complex as human brains). However, organizing this computing power to learn as generally as the human brain does will be a real challenge and add years or decades to the length of time required to create intelligent machines.

Dreyfus stops short of saying that intelligent machines are impossible, but rather argues that current approaches will not work for the most difficult intelligent behaviors. He makes it clear that he is reacting to overly optimistic claims by machine intelligence researchers. This is a valuable insight, but it is important to use it to redirect rather than abandon research.

REASONING FROM IMPOSSIBLE HYPOTHESES

It is interesting that both Penrose's argument against weak artificial intelligence and Searle's argument against strong artificial intelligence reason from impossible hypotheses. Penrose's argument uses Turing machines, which have infinite tapes that cannot exist in our finite universe.

Searle's Chinese Room argument uses a lookup table that is finite but contains many more entries than there are particles in the universe.

These impossible hypotheses seem to me like modern scientific analogs of the non-physical explanations of consciousness from ancient and modern religions. The ultimate explanation of consciousness will probably use our current "mundane" understanding of neurons, made special only by the incredible complexity of interactions among the 100 billion neurons and 100 trillion synapses in the human brain. These numbers are large but possible. Penrose's and Searle's arguments get into trouble in the realm of impossibly large numbers. On the other hand, we can see Dreyfus's argument as saying that research on machine intelligence goes astray when it tries solutions whose complexity is much smaller than the numbers of human neurons and synapses.

NOTES

1. Penrose, 1994.
2. McCullough and Pitts, 1943. Hopcroft and Ullman, 1969.
3. Turing, 1936. Hopcroft and Ullman, 1969.
4. Tipler, 1994.
5. Winograd, 1972.
6. Lester and Porter, 1997.
7. Searle, 1984.
8. Dreyfus, 1979.
9. Haeckel, 1900. http://www.ucmp.berkeley.edu/history/haeckel.html.

Chapter 5

THE CURRENT STATE OF THE ART IN MACHINE INTELLIGENCE

Computer scientists are trying to create computer systems that exhibit intelligent behavior, in a field called *artificial intelligence* or simply AI. Some researchers don't worry whether their computer systems work in the same way that human brains do. For example, most chess playing programs examine much larger numbers of possible lines of play than humans can, but cannot explain their plans for defense and attack in the way that a human chess player can. Other researchers specifically try to mimic the way human brains work. For example, there have been efforts to mimic conscious human reasoning in programs for proving mathematical theorems and even in programs for game playing. Hubert Dreyfus's criticism was directed at such efforts that model higher-level reasoning without any model for lower-level, unconscious brain processes. Recognizing the validity of his criticism, some researchers have recently made efforts to mimic the behaviors of simpler animals without any claim to intelligence such as multi-legged robots that mimic the way insects walk.[1]

Early AI researchers made overly optimistic predictions of when their computer systems would exhibit intelligent behavior comparable to humans, some even predicting that such machines would exist by the year 2000. The failure of these predictions has triggered a backlash, including the arguments discussed in Chapter 4. This chapter is a description and assessment of current AI work, and an assessment of future prospects.

GAME PLAYING

The AI field had an early focus on creating machines that could beat humans at mental games such as chess and checkers. This is because ability at these games is considered an indicator of intelligence among humans, and because there are relatively simple programming techniques for creating effective game playing computers. Rather than mimicking human reasoning about games, computers can use their computational speed to examine a very large number of future lines of play to calculate their next move. As early as 1959 a program using these techniques was able to beat a human checkers champion.[2]

To understand the basic computer technique for game playing, consider the problem of programming a computer to play tic tac toe. There are two players X and O, and we can label the board positions as:

	1	2	3
A	A_1	A_2	A_3
B	B_1	B_2	B_3
C	C_1	C_2	C_3

As an example, imagine that the computer is player X and the current board position is:

	1	2	3
A	X		
B	O		X
C	X	O	O

It is X's turn to play and it has three possible choices: A_2, B_2 and A_3. If X chooses A_2, then O has two possible choices: B_2 and A_3. If X chooses B_2, then O has two possible choices: A_2 and A_3. And if X chooses A_3, then O has two possible choices: A_2 and B_2. In every case, after O's response X has only one choice left and the outcome is determined as one of: X wins, O wins or draw. We can lay all these possible lines of play out in a hierarchical outline:

X's play	O's play	X's play	outcome
A_2-X			
	B_2-O		
		A_3-X	X wins
	A_3-O		
		B_2-X	draw
B_2-X			
	A_2-O		
		A_3-X	X wins
	A_3-O		
		A_2-X	draw
A_3-X			
	A_2-O		
		B_2-X	X wins
	B_2-O		
		A_2-X	X wins

From this hierarchy it is clear that X should choose A_3, since this gives X a win no matter which choice O takes. If X chooses either A_2 or B_2, O has a choice that forces a draw.

Here the hierarchy is quite simple. There are only 3 empty board locations and hence only $3 \times 2 \times 1 = 3! = 6$ possible future lines of play. At the start of a game there are 9 empty board locations and hence $9! = 362880$ possible future lines of play. Searching all these lines of play is difficult or impossible for a human but easy for a computer.

And a program for searching all possible future lines of play in tic tac toe is not too complex. The program chooses the best move using what is called a min-max search. This assigns the values to final board positions as +1 for a win by X, 0 for a draw, and -1 for a win by O. Then it works backward from the final positions. If it is X's turn to play in a board position, then the value of that position is the maximum of the values of all possible next board positions, assuming X will make its best move. And if it is O's turn to play in a board position, then the value of that position is the minimum of the values of all possible next board positions, assuming O will make its best move. By working backward from final board positions to the current position, this procedure assigns values to all possible future board positions. Faced with a current board position, the program has only to choose the play that leads to the next position with the best value.

This works for games like tic tac toe that can be exhaustively searched, but for a more complex game like chess there are too many future lines of play for even the fastest computer to search them all. So a chess-playing computer cannot follow each line of play to the end of the game to see which player wins. In order to work backward in lines of play using min-max search, the program must estimate the values of board positions that are not final (i.e., where more play is possible). In chess the values of board positions may be estimated, not very accurately, by counting how many pieces each player has left. Since a queen is more valuable than a pawn, the program can improve the accuracy of this estimate by giving different values to queens, pawns and other pieces. But how can the program determine the relative value of the different pieces? To answer this, assume we have a number of different theories about what the relative values of pieces should be. We can hold a tournament among a set of chess playing programs that use these different theories about piece values but are otherwise identical. The winner of the tournament will have the best set of values for the pieces.

But rather than holding a single tournament with a large number of programs, it is more efficient to hold a series of matches between pairs of programs. After each match the winner is the new champion, and a challenger is chosen with a slightly different set of relative piece values than the champion. In fact, a good chess program will evaluate board positions based on more than just values of pieces, so there will be many different numerical parameters to adjust in order to get the best program. A series of matches to determine optimum parameters is a form of machine learning. Numerical factors can also be adjusted by comparing a program's choices with those made by human chess champions in the board positions of their games.

For large min-max searches there is a technique called alpha-beta pruning that can greatly reduce the number of possible future lines of play that have to be searched. Imagine again that the computer is player X in tic tac toe and is trying to choose the best move in a given board position. Say the computer has already completed its search for one possible move M_1 and found that it can get at least a draw. Now it is searching all possible O responses to another move M_2, and finds one O response where O is guaranteed to win. Then there is no need to search any other O responses to move M_2, since the draw for move M_1 is a better choice for X than the loss for move M_2. These other O responses are "pruned" from the search. Alpha-beta pruning can be applied at every stage of the search, reducing the size of the search to a certain fraction at each stage. For deep searches with many stages, the product of these fractions results in a tiny fraction of the tree that actually needs to be searched.

IBM's Deep Blue chess machine used alpha-beta pruning plus parallel computers that could search hundreds of millions of future lines of play on each move.[3] This allowed very deep searches. The Deep Blue team worked for years to create accurate evaluations of board positions. They also developed techniques for detecting important lines of play and increasing the depth of search along those lines. All of this enabled Deep Blue to beat the world chess champion, Gary Kasparov. No one involved thought that Deep Blue was intelligent, because its min-max search cannot generally be applied to other intelligent behaviors. It is different from the way human chess players think. Hence game playing machines like Deep Blue do not provide much guidance about how to build truly intelligent machines. However, as previously mentioned, Kasparov had a strong emotional feeling that Deep Blue was thinking. And that does tell us something about the emotional connection that will develop between people and intelligent machines.

EXPERT SYSTEMS

Expert systems are intended to use certain techniques taken from AI research in order to produce systems that can provide practical expert advice on specialized topics. The AI techniques include:

1. Representing complex knowledge.
2. Reasoning from that knowledge.
3. Understanding questions and generating answers in natural human language (some expert systems employ only minimal language abilities).

To illustrate by a simple example, we define a knowledge-base (i.e., a knowledge database) that contains only statements of the form "X is-a Y." Our specific knowledge-base might consist of the statements:

Tabby is-a horse
Sting is-a horse
Sarah is-a dog
horse is-a animal
dog is-a animal

We can define a reasoning engine with only one rule of deduction:

if "X is-a Y" and "Y is-a Z" then "X is-a Z"

And we can define a simple natural language grammar:

question := Is X a Y?
answer := yes
answer := no

If we ask "Is Tabby a animal?" the language parser will match this question with X = Tabby and Y = animal. The match to the question will trigger the reasoning engine to try to prove "Tabby is-a animal." Even a simple exhaustive search strategy in the reasoning engine will find "Tabby is-a horse" and "horse is-a animal" and so prove "Tabby is-a animal." This would then trigger the system to generate the answer "yes."

While this example is simple, it illustrates the basic idea of expert systems. Practical expert systems have much larger knowledge-bases and more complex rules for reasoning. In such systems the ability to answer questions depends on efficient search strategies in the reasoning engines to find long sequences of deductions. Some expert systems work with probabilities rather than simple true or false values, and perform probability computations in their reasoning engines.

An excellent early example was the MYCIN system for medical diagnosis.[4] This system diagnosed infections and recommended antimicrobial treatments based on descriptions of patient symptoms. Like many expert systems, MYCIN asked the user a series of questions with simple answers, so that it did not need to interpret complex user statements written in natural human language. Thus a MYCIN session might start with the typed conversation between the computer and the user something like:

Computer: Patient's name:
User: John Smith

Computer:	Sex:
User:	M
Computer:	Age:
User:	63
Computer:	Have any organisms been grown from cultures relevant to John Smith's illness?
User:	Y
Computer:	Where was the culture specimen taken?
User:	Blood
Computer:	What are the date and time when the specimen was taken?
User:	February 14, 1998
Computer:	Identity of the first organism grown from the culture.
User:	Unknown

As the user answers questions, the system reasons based on those answers in order to formulate new questions. For example, when the user says that the identity of the culture organism is unknown, the system will ask questions to try to determine its identity. The answers to these questions are converted into logical statements that are then combined with the system's knowledge-base by the system's reasoning engine. The knowledge-base includes several types of rules. There are rules for identifying organisms, for example:

If the organism has a gramneg stain, and
    has rod morphology, and
    is aerobic
Then there is probability 0.8 that the organism is
    enterobacteriaceae

There are other rules for recommending therapy, for example:

If the organism is bacteroides
Then recommend one of the following drugs:
    Clindamycin with probability 0.99
    Chloramphenicol with probability 0.99
    Erthromycin with probability 0.57
    Tetracycline with probability 0.28
    Carbenicillin with probability 0.27

Because there is so much uncertainty in medicine, MYCIN's reasoning engine made probability computations both for diagnosing infections and for recommending therapies. Where multiple rules apply to therapy recommendation, the system combines probabilities to determine the therapy that is most probably correct. MYCIN performed as well as many physicians but was never used in practice because of legal and ethical issues. Furthermore, it was difficult to keep the system's knowledge-base up to date with current medical knowledge.

There are examples of expert systems in practical use. The AskJeeves web search engine accepts a user question in English, interprets it using its knowledge-base and reasoning engine, and produces a list of web addresses that should be useful for answering the question.[5] The R1/XCON system was used commercially to help configure VAX computer systems. R1/XCON is just one example of an expert system application to help sales representatives avoid errors in specifying complex products for customers.

The primary problems with expert systems are that their knowledge-bases have limited scopes, and are difficult and expensive to create and maintain. For example, a medical expert system may not include any rule to abandon treatment if the patient has died. This is certainly obvious to any human, and of course no human would employ an expert system to recommend treatment for a dead patient. But it is an example of the way that expert systems can lack common sense.

Some people believe that the solution to the common sense problem lies in creating very large knowledge-bases that include or imply everything that humans know about the world. One approach to building large knowledge-bases is to develop systems that increase the efficiency by which human experts enter their knowledge. Another approach is finding ways to automate their construction by giving computers the ability to learn rules on their own.

Hubert Dreyfus argues that the general expert systems approach using discrete facts in knowledge-bases and reasoning engines must fail to create anything like human intelligence, and I agree with him. First, it is clear that human brains do not contain anything analogous to these knowledge-bases. Human brains remember by recreating sensory experiences, which are very rich in detail, although not always accurate, and exist in very large number. Even a discrete fact like "two plus two equals four" is remembered as the visual or auditory experience of its statement. And no set of language statements can capture the richness of our memories of a person or animal we know well. Second, seemingly discrete facts can be picked apart until they seem to lose all meaning. Examples include the philosophy teacher's exercise of challenging students to precisely define the difference between the words "table" and "chair," and the president's defense based on questioning the meaning of the word "is." Such exercises work because human experiences with chairs and tables always exceed our capacity to encode them in language.

KNOWLEDGE REPRESENTATION

In a sense, any data stored in a computer is some form of knowledge representation. However, AI research has developed special forms of knowledge representation that reflect various theories of the way that brains work, and that are useful for solving the challenging behavioral problems that AI addresses. These include:

1. Statements in some logical calculus.
2. Semantic networks of entities and relations among those entities.
3. Frames, which are like objects in an object-oriented programming language, connected together in a network.
4. Procedures which define the actions of items of knowledge.
5. Mathematical models of the physical world, used for example by robots that must navigate in a world governed by the laws of physics.
6. Probabilities learned from experience with the world.
7. Features for vision and other pattern recognition applications.
8. Abstract networks and connection strengths, in a connectionist system modeling the way that human and animal neurons work.

The expert system examples in the previous section used logical statements to represent knowledge. These can be formalized as formulas in some logical calculus. The simplest example is the *propositional calculus*, which consists of an alphabet A of atomic formulas (e.g., the symbol $a \in A$ may represent "Sarah is good"), plus the connective symbols: $\wedge$ (and), $\vee$ (or), $\neg$ (not), $\Rightarrow$ (implies) and parentheses. Valid propositional formulas are defined by:

1. Any $a \in A$ is a formula.

2. If x and y are formulas, then:

$\neg x$

$(x \wedge y)$

$(x \vee y)$

$(x \Rightarrow y)$

are all formulas.

The propositional calculus can be used to represent knowledge like ("Sarah chews the furniture" $\Rightarrow$ "Sarah is bad"). But we know that anyone who chews the furniture is bad, and we need a way to express this more general knowledge rather than having many propositional formulas of the form ("x chews the furniture" $\Rightarrow$ "x is bad").

For this we need the *predicate calculus*, which is the propositional calculus plus an alphabet V of variables (like x), an alphabet C of constants (like Sarah), an alphabet P of predicates, and the quantifiers $\forall$ (for all) and $\exists$ (there exists). Constants represent specific individuals like Sarah, while variables can represent any individual. Predicates represent statements about zero or more individuals (note that predicates about zero individuals play the role of atomic propositional formulas). For example, the predicate $f(x)$ may represent the statement "x chews the furniture" and the predicate $g(x)$ may represent the statement "x is bad." Then our general knowledge may be represented by the predicate calculus formula:

$(\forall x)(f(x) \Rightarrow g(x))$

This can be read as "for all x, x chews the furniture implies x is bad." Valid predicate calculus formulas are defined by:

1. Any predicate $p \in P$, applied to an appropriate number of constants and variables, is an atomic formula.

2. If x and y are formulas, then:

$\neg x$

$(x \wedge y)$

$(x \vee y)$

$(x \Rightarrow y)$

are all formulas.

3. If x is a formula and $v \in V$, then:
    $(\forall v)x$
    $(\exists v)x$
    are formulas.

What I have described is the first order predicate calculus. Second order predicate calculus includes variables that represent first order predicates, and second order predicates that represent statements about zero or more individuals and first order predicates. However, most logical knowledge representations use only first order predicate calculus.

Knowledge representations are often restricted to *Horn clauses*, which are first order predicate calculus formulas of the form:

$$a_1 \wedge a_2 \wedge \ldots \wedge a_n \Rightarrow b$$

where a_1, a_2, ..., a_n and b are atomic formulas. The PROLOG programming language is based on Horn clauses. A PROLOG program is basically a collection of Horn clauses that express some knowledge, and its input is a set of atomic formulas to all be proved true. PROLOG is widely used by AI researchers, and was the basis of the Japanese Fifth Generation computing project.[6] One great advantage of PROLOG is that a running program is basically a search for a proof and such searches allow lots of parallelism with different processors trying different solutions. The Fifth Generation project tried to exploit Japanese expertise in designing parallel computers to create a significant advance in practical applications of AI. While the project was a technical success, it suffered from the general failure of the rule-based expert system approach to achieve intelligence.

There are several active AI projects trying to create enormous knowledge-bases represented as logic formulas. Doug Lenat's Cycorp has developed a knowledge-base containing hundreds of thousands of rules.[7] Their system, named Cyc and the basis of numerous commercial applications, represents knowledge using first order predicate calculus with extensions for certain specialized knowledge (e.g., some limited second order formulas, and the theory of equality which cannot be expressed in first order logic). The Cyc effort is based on the assumption that systems can behave with common sense if they have sufficiently large knowledge-bases and reasoning engines capable of handling them. While I am pessimistic about the prospects for intelligence from rule-based systems, efforts involving very large knowledge-bases deserve an open mind about what they will achieve.

Semantic networks represent knowledge by labeled nodes, which represent entities, and labeled directional arcs between pairs of nodes, which represent binary relations between entities. For example, a semantic network may contain nodes labeled *Tom* and *Dick*, and an arc labeled *father* from *Tom* to *Dick*. Semantic networks are equivalent to predicate calculus without variables and with only binary predicates. Their advantage is that they lend themselves to graphical depictions that are easy for people to visualize. Some AI researchers have experimented with ways to extend semantic networks, but they are all equivalent to logical representations.

Frames are much like objects in an object-oriented programming language. A frame has a number of slots, each with its own type. A slot may hold a reference to another frame, or may hold a list of values. Because of these references between frames, frames are sometimes classified as an extension of semantic networks. Like objects, frames have classes with inheritance relations among classes of frames. For example, a class of frames for *dogs* may inherit from a class of frames for *mammals. Mammal* frames may include a slot for *weight*, which would be inherited by *dog* frames. *Dog* frames may add a slot for *breed*, which may itself be a class of frames.

The Biology Knowledge-Base developed by AI researchers at the University of Texas contains over 22,000 frames.[8] A typical frame in this knowledge-base defines *snowmelting* with a frame:

Frame name: *snowmelting*
Slot name: *transformed_entity*   slot value: H_2O
Slot name: *after_state*   slot value: *water*
Slot name: *before_state*   slot value: *snow*
Slot name: *generalizations*   slot value: *melting*

Note that the slot values in this frame are references to other frames. The Biology Knowledge-Base is used for a variety of advanced AI research projects, such the KNIGHT and TRIPEL systems described in later sections.

Procedural knowledge representations come in a variety of forms. A frame representation that includes slots for procedures is essentially an object-oriented programming language. Procedures are sometimes linked to patterns in a logical knowledge-base, so that whenever the system finds a set of logical statements matching a pattern the corresponding procedure is invoked. There is also the classic example of procedural knowledge in Terry Winograd's SHRDLU system.[9] It expressed knowledge of English grammar as a set of procedures. For example, the procedure for *sentence* invoked the procedure for *noun-phrase* and the procedure for *verb-phrase*. These procedures searched for matches with the words of a sentence. One advantage is that the procedures for syntactic structures can invoke procedures to check semantic consistency. That is, procedural representations naturally support linkages between different types of knowledge.

Mathematical models are used to represent knowledge about the physical world, which is useful for robots that need to interact with physical objects. One fascinating example is an annual soccer tournament played between teams of robots designed by teams of students from universities with leading AI research programs.[10] These soccer playing robots need models of how the soccer ball moves and how they move. Sophisticated soccer robots will try to learn a mathematical model of how the robots on the opposing teams move.

Probability information is often added to logical knowledge representations, to extend the simple choice of true or false with numerical probabilities of truth. This can be any value between 0.0 (false) and 1.0 (true). There is also fuzzy set theory, in which the statement of set membership $a \in A$ is assigned a value between 0.0 and 1.0. With such knowledge representations, the reasoning engine includes probability calculations.

Artificial vision systems require knowledge in the form of visual features for recognizing objects and understanding scenes. For example, systems used by the U.S. Post Office to read addresses on envelopes try to isolate individual characters on the envelope and match them against a knowledge-base of features for letters and numbers. Similarly, speech recognition systems used for some automated telephone response systems match spoken words against a knowledge-base of features for the basic sounds of the English language.

All of the knowledge representations we have considered so far can be understood by humans. In fact, they are usually explicitly created by humans. On the other hand, knowledge representations in human and animal brains, even in simple animal brains, defy human understanding. Knowledge in human and animal brains is represented by the topology of their neural networks, and by the behavioral attributes of individual neurons and connections. If you examined my brain you could never find the representation of my knowledge that "any dog that chews the furniture is a bad dog." It is hopelessly distributed over many neurons and their connections. Even artificial neural networks, constructed by humans, represent knowledge in ways that are similarly difficult or impossible for humans to understand.

Some sort of neural network knowledge representation is probably necessary to achieve real machine intelligence, because of the terrific adaptability required for intelligence. That is, any knowledge representation intelligible to a human will be limited by the specific experience and education of that human, and not be as adaptable to utterly foreign situations as a human brain is. On the other hand, human and animal brains include large numbers of specialized behaviors that may be more efficiently implemented using more explicit knowledge representations. For example, physical phenomena like weather can be modeled much more efficiently by numerical weather models than by neural networks.

REASONING

Just as any data in a computer can be interpreted as a form of knowledge representation, any program execution can be interpreted as a form of reasoning. However, AI research has developed special forms of reasoning that are useful for solving the challenging behavioral problems that AI addresses.

Many AI systems reason by searching for proofs in predicate calculus. Knowledge represented as semantic networks or frames can be phrased in terms of predicate calculus for reasoning. The appeal of this approach is that virtually any explicit human knowledge can be expressed in terms of some form of predicate calculus, and a very general proof procedure, called *resolution*, exists for predicate calculus.[11]

Resolution works with formulas in a special form called *clause form*, and there is a cookbook set of steps for converting any predicate calculus formula into clause form. It is easiest to understand by an example (note in this example x and y are variables, a is a constant, and f, g and h are predicates):

$$((\forall x)(\exists y)((\neg f(y, a) \wedge g(x, y)) \Rightarrow h(x)) \wedge (\forall x)k(x))$$

Step 1. Eliminate $\Rightarrow$ by the equivalence $(a \Rightarrow b) \equiv (\neg a \vee b)$:

$$((\forall x)(\exists y)(\neg(\neg f(y, a) \wedge g(x, y)) \vee h(x)) \wedge (\forall x)k(x))$$

Step 2. Move $\neg$ inward according to various logical rules, such as $(\neg(\neg P) \equiv P$ and $\neg(P \wedge Q) \equiv (\neg P \vee \neg Q)$:

$$((\forall x)(\exists y)((f(y, a) \vee \neg g(x, y)) \vee h(x)) \wedge (\forall x)k(x))$$

Step 3. Rename variables so each quantifier uses a unique name:

$$((\forall x)(\exists y)((f(y, a) \vee \neg g(x, y)) \vee h(x)) \wedge (\forall z)k(z))$$

Step 4. Move all quantifiers to the left but don't change their relative order:

$$(\forall x)(\exists y)(\forall z)((f(y, a) \vee \neg g(x, y) \vee h(x)) \wedge k(z))$$

Step 5. Eliminate all existential quantifiers (e.g., $\exists y$) by replacing their variables by functions of the universally quantified variables (e.g., $\forall x$) to their left (e.g., replace y by $q(x)$):

$$(\forall x)(\forall z)((f(q(x), a) \vee \neg g(x, q(x)) \vee h(x)) \wedge k(z))$$

Step 6. Drop universal quantifiers, which are implicit:

$$((f(q(x), a) \vee \neg g(x, q(x)) \vee h(x)) \wedge k(z))$$

Step 7. Convert to conjunctive normal form, which means that at the top level the formula is an and-expression ($\wedge$) of a series of sub-formulas, each of which is an or-expression ($\vee$) of atomic formulas or negations of atomic formulas. Note that our example is already in this form:

$$((f(q(x), a) \vee \neg g(x, q(x)) \vee h(x)) \quad \wedge \quad k(z))$$

Each sub-formula in the top-level and-expression is a clause:

$$(f(q(x), a) \vee \neg g(x, q(x)) \vee h(x))$$
$$k(z)$$

As a final step, if any variable name occurred in more than one of these clauses, it would be changed to different names in each clause.

In order to use the resolution proof procedure, the entire knowledge-base has to be in clause form. Then any formula to be proved is negated, put in clause form, and added to the other clauses in the knowledge-base. The goal of the proof procedure is to derive a contradiction (an empty clause), which proves that the negation of the original formula must be false, and hence that formula must be true. The actual resolution procedure simply searches for pairs of clauses where one contains an atomic formula and the other contains its negation, and combines them minus the formula and its negation. In order to match a pair of clauses, it may be necessary to substitute variables, constants or functions from one clause for variables in the other clause.

For example, recall the simple knowledge-base used in a previous section to illustrate expert systems. Here we express the knowledge in clause form with a predicate $is\text{-}a(x, y)$ which means "x is-a y":

> $is\text{-}a$(Tabby, horse)
> $is\text{-}a$(Sting, horse)
> $is\text{-}a$(Sarah, dog)
> $is\text{-}a$(horse, animal)
> $is\text{-}a$(dog, animal)

The rule of deduction used with this example was:

> if "X is-a Y" and "Y is-a Z" then "X is-a Z"

This is no longer a rule of deduction; that role is taken by resolution. Rather, it is now expressed as knowledge in clause form:

> $\neg is\text{-}a(x, y) \lor \neg is\text{-}a(y, z) \lor is\text{-}a(x, z)$

The question posed in this example was "Is Tabby an animal?" Resolution works with the negation of this in clause form:

> $\neg is\text{-}a$(Tabby, animal)

The first resolution step is to merge this with the clause expressing the rule of deduction, substituting Tabby for x and animal for z. The result is:

$\neg is\text{-}a(\text{Tabby}, y) \lor \neg is\text{-}a(y, \text{animal})$

This is then merged with $is\text{-}a(\text{Tabby}, \text{horse})$, substituting horse for y. The result is:

$\neg is\text{-}a(\text{horse}, \text{animal})$

Finally this is merged with $is\text{-}a(\text{horse}, \text{animal})$ resulting in the empty clause, hence proving the truth of the question "Is Tabby an animal?"

The resolution procedure is widely used by AI systems, and they employ a variety of search strategies for deciding the order in which to try pairs of clauses. The strength of the resolution procedure is its generality: it can be applied in almost any AI application. Its weakness is that it does not make use of specific application knowledge that can guide a more efficient search.

A very different style of reasoning is used in systems with procedural and model knowledge representations. A procedural representation of knowledge can define not only the conditions under which something is true, but also the sequence of operations for proving it. For example, the $is\text{-}a$ predicate used to illustrate expert systems had the rule of inference:

$$is\text{-}a(x, y) \land is\text{-}a(y, z) \Rightarrow is\text{-}a(x, z)$$

This means that the $is\text{-}a$ predicate is a transitive binary relation. These have been extensively studied, and efficient procedures have been developed for converting a knowledge-base of $is\text{-}a(a, b)$ statements into a data structure that allows quick look-up of $is\text{-}a(x, y)$ for any x and y. We can use one of these efficient procedures for answering questions of the form "Is X a Y?"

The TRIPEL system for answering biology questions, developed at the University of Texas, is a fascinating combination of logical and model-based reasoning.[12] It uses the Biology Knowledge-Base discussed in the previous section, and predicate calculus reasoning, to construct mathematical models of plant processes. Based on its knowledge-base, it determines which processes work at relevant time scales and levels of detail in order to construct a reasonable differential equation model of the process in question. It also applies its knowledge to construct reasonable boundary and initial conditions for the equations. TRIPEL solves the resulting differential equations numerically and uses the solution to formulate an answer.

Robots also reason using mathematical models of physical objects, including locations of walls and other obstacles, locations and speeds of objects of interest and of the robots themselves. Their mathematical reasoning must encode the knowledge that two objects cannot occupy the same space or move through each other, as well as knowledge of the way the robot will move in response to actions of the robot's legs or wheels. While reasoning about physical systems could be accomplished using logical reasoning, this would be much less efficient than numerical mathematical models.

Reasoning for robots and other systems must manage knowledge-bases that change over time, especially in response to the robot's own actions. That is, complex robot actions must be formulated as plans that involve a sequence of changing situations. A classic example of planning was used in the SHRDLU system for natural language. The subject of its conversation was a set of blocks of various shapes, sizes and colors that could be stacked (they were not physical, but existed only on a display screen). The user could ask the system to stack the blocks in various ways. In a planning system, actions have preconditions that must be satisfied before the action can occur, and postconditions that will true after the action occurs. For example, given blocks x and y, actions might be:

Preconditions: $clear(x)$, $clear(y)$
Action: $stack(x, y)$
Postconditions: $on(x, y)$, $\neg clear(y)$

Preconditions: $on(x, y)$, $clear(x)$
Action: $remove(x, y)$
Postconditions: $\neg on(x, y)$, $clear(y)$

Note the postconditions only define changes to predicate values, so $clear(x)$ is not specified in the postconditions of these actions because it is part of their preconditions. A sequence of actions defines a sequence of knowledge-bases. Since each is slightly different from the previous knowledge-base, there has been much work on developing efficient representations for sequences of knowledge-bases that do not duplicate knowledge that does not change over the sequence. The knowledge-base should also include general knowledge such as:

$(\forall y)(clear(y) \Leftrightarrow (\forall x)\neg on(x, y))$

This says that a block is clear if and only if there is no block on it.

In ordinary resolution reasoning the system searches for a sequence of merge operations on the knowledge-base that result in an empty clause. In planning reasoning, the system searches for a sequence of actions that result in a knowledge-base with certain conditions true. For example, asking the system to create a stack of blocks a, b and c is expressed as asking for a sequence of actions that make $on(b, a)$ and $on(c, b)$ both true.

Machine reasoning systems have demonstrated abilities that rival human reasoning in narrow mathematical specialties. Some AI researchers believe that current machine reasoning systems are sufficient to match human capabilities, if they could be combined with knowledge-bases as large and broad as human knowledge. I am skeptical but open minded about this. In human and animal brains, knowledge and reasoning are factored very differently from most AI systems. Knowledge and reasoning are both encoded in the topology of the neural network and attributes of neurons and connections, in ways that are impossible for human observers to disentangle.

NATURAL HUMAN LANGUAGE

SHRDLU is the classic system for understanding natural language. This system had a very small knowledge-base about a world of blocks of various shapes, sizes and colors that could be stacked. But it could understand complex English sentences about its block world. Here is an example of a conversation with SHRDLU:

User:	Pick up a small blue pyramid.
SHRDLU:	OK
	(the display shows the small blue pyramid rising)
User:	Put it on the red block.
SHRDLU:	I don't understand which red block you mean.
User:	Find the pyramid that is shorter than the one you are holding and put it into the green box.
SHRDLU:	OK
	(the display shows the action happening)
User:	What is sitting on the large red block?
SHRDLU:	The small green block.

SHRDLU encoded knowledge and reasoning procedurally in a programming language called PLANNER, which was based on LISP. It encoded its English grammar in another LISP-based language called PROGRAMMER. Natural language is full of ambiguity. For example, the sentence "Time flies" can either be a statement that time passes quickly, or a command to measure the times of house flies. These meanings correspond to parsing "Time flies" as either noun-verb or verb-noun. A key problem for a natural language understanding system is to resolve such ambiguous parsings. In SHRDLU, the parsing procedures invoked the semantic procedures to immediately verify whether an attempted parsing made sense. That is, syntax and semantics were interacting rather than sequential processes. SHRDLU did not try to understand spoken language, but in systems that do it is equally important that the process of recognizing words from sounds interact with syntax and semantic processes. The only accurate way to resolve the ambiguities of low level language processes is by interaction with higher level processes. For example, if a companion has been talking about the growth of his children and then says "Time flies," we know from the context that he is making a philosophical observation about how quickly time passes, rather than commanding us to take out a stop watch and measure the activities of flies.

The high quality of SHRDLU's conversations convinced many AI researchers that the system had solved the basic problems of machine natural language, and that all that was required was giving machines much larger knowledge-bases than SHRDLU's simple blocks world. The Biology Knowledge-Base at the University of Texas containing over 100,000 facts is the basis for several natural language experiments. One is the KNIGHT system for generating natural language explanations for biological phenomena.[13] The explanations of this systems were given A-F letter grades by a panel of biology experts, in competition with explanations written by a second panel of biology experts. KNIGHT scored within one-half letter grade of the second expert panel, and actually outscored one expert.

If you want to try a conversation with a high-quality AI language system, you can talk to the START system at MIT.[14] This system uses a knowledge-base that covers a variety of topics, including MIT and AI. Here is a conversation with START that demonstrates its considerable language abilities, as well as the limits of its knowledge:

User:	Where is MIT?
START:	Massachusetts Institute of Technology

	77 Massachusetts Avenue, Room 5-111, News Office
	Cambridge MA 02139
	(617) 253-1000
	Admissions E-mail: admissions@mit.edu
	Web site: http://web.mit.edu/admissions/www
User:	How many students at MIT?
START:	Massachusetts Institute of Technology

4,258 (males and females)
Source: U.S.News
The MIT Artificial Intelligence Laboratory's 178 members include 17 faculty members, 26 academic staff, 29 research and support staff, and 106 graduate students.

User: Is MIT a good school?

START: Yes...
           START believes that MIT is the best university in the world.

User: Is Harvard a good school?

START: Unfortunately, I wasn't told if Harvard is a good school.

User: Is MIT a good place to get a sandwich?

START: Sorry - I don't know whether or not MIT is a good place to get a sandwich.

User: Who was President of the United States in 1915?

START: Woodrow Wilson: March 4, 1913 to March 3, 1921

User: What were some major achievements of President Woodrow Wilson?

START: I don't have this information.

User: Who was Woodrow Wilson's Vice President?

START: Woodrow Wilson's Vice President was Thomas R. Marshall (1913-21).

User: Are computers intelligent?

START: I don't know the answer to your question.

User: Will computers be as intelligent as humans?

START: I'm not sure I understand what you are saying. Could you try it again?

This conversation demonstrates that START does a good job of understanding questions but is weak in the small size of its knowledge compared to human knowledge.

One practical application of AI research in natural language understanding is automated translation. There are a number of on-line services for translating text or web pages from one language to another.[15] While the translations are not perfect, they usually do a reasonable job of communicating the basic ideas.

Understanding spoken or hand-written speech adds difficult pattern recognition problems to language understanding. However, significant progress is being made in both areas through machine learning. Spoken speech recognition has progressed to the point where it is used for some commercial telephone answering systems, and IBM even has a commercial product for this purpose.[16] Carnegie Mellon University has a long term speech recognition research project called Sphinx that supplies the results of its research as free open-source software.[17] The Sphinx project even provides a toll free telephone number where you can try their system for making flight reservations. It performs reasonably well. The big problem for recognizing spoken language is separating a voice from background noise. Current systems could never carry on a conversation at a loud party.

AI research is making sufficient progress with language understanding to support real applications. The major deficiency is in the knowledge needed to understand the meaning of language. Imagine that a person and a computer share a common human friend. The person will have many memories of the way the friend looks, moves, sounds and thinks. The computer's knowledge-base will include a set of facts about the friend but not come close to the richness of the person's memories. This difference would become apparent in any long conversation about the common friend.

MACHINE VISION

Vision must process much more raw data than other senses, and hence requires the greatest processing power. And because vision sees objects in a geometrical space measured by numerical distances and coordinates, and in color intensities with numerical values, computer vision makes greater use of numerical mathematical algorithms than other AI fields. The input to a basic vision system consists of a time sequence of 2-D images (containing of course depictions of 3-D objects).

Images are formed by light reflected, or occasionally emitted, from 3-D objects. Light consists of a distribution of wavelengths that we perceive as color. For example the distribution from our sun defines the color we know as white. The intensity and distribution change when light reflects from an object. The change in distribution gives objects their apparent color. Intensity depends on the angle of the light striking an object's surface, and variations in apparent intensity along a surface give us the sense of the object's shape. Images are formed by light reflected from many objects, and some of these objects block our view of other objects. Other complexities in image formation arise from light emitted directly from objects such as light bulbs, light absorbed by atmospheric phenomena such as fog, light refracted by transparent objects such as lenses, and light reflected by shiny objects such as mirrors. Finally, light entering an eye or computer camera is sampled at a finite number of detectors over a 2-D retina or focal plane, which respond to a finite number of wavelength bands.

The goal of a vision system is to reverse this complex image formation process, and deduce 3-D geometry descriptions and identifications of the objects and atmospheric conditions that formed the image. As with natural language, artificial vision systems divide into several interacting processes.[18] The lowest level processes deal with raw images from computer cameras, which generally consist of a rectangular grid of pixels with numerical intensities for either one (gray level) or three (red, green and blue) wavelength bands. Inside a computer a color image is typically stored in numerical arrays:

    float red[nlines][nelements]
    float green[nlines][nelements]
    float blue[nlines][nelements]

Low level vision processes partition the image into distinct regions, which may come from different 3-D objects. This is often done by *edge detection* algorithms, which look for region boundaries as abrupt color or intensity changes. These algorithms look for large differences between adjacent pixels. These may be simple differences like:

    red(i+1, j) - red(i−1, j)
    red(i, j+1) - red(i, j−1)

However, such simple operators are susceptible to image noise. Thus average differences are usually computed over numbers of pixels, such as:

$$((\text{red}(i+1, j+1) + \text{red}(i+1, j) + \text{red}(i+1, j-1)) -$$
$$(\text{red}(i-1, j+1) + \text{red}(i-1, j) + \text{red}(i-1, j-1))) / 3$$
$$((\text{red}(i+1, j+1) + \text{red}(i, j+1) + \text{red}(i-1, j+1)) -$$
$$(\text{red}(i+1, j-1) + \text{red}(i, j-1) + \text{red}(i-1, j-1))) / 3$$

Pixels with large differences for red, green or blue values are marked as edges. Differences may be normalized according to average intensities over small image regions. Because these differences are spatial vector quantities (i.e., they have i and j components) they define edge strength and edge direction for all pixels, with edge pixels identified as those with strength over a certain threshold.

Edge pixels detected by such simple difference operators will generally not form continuous region boundaries. A variety of algorithms are used to connect edge fragments into continuous boundaries, and to discard edges that do not connect as parts of boundaries. Edge relaxation algorithms start with an edge pixel and increase the edge strengths of nearby pixels that would fall along a common edge, as determined by the edge direction of the edge pixel. Any pixels not strengthened in this way have their edge strengths reduced. Isolated edge pixels loss strength and cease to be edge pixels, while continuous edges broken by isolated missing edge pixels tend to fill in. Iterating this process can create continuous curves of edge pixels. An edge can also be propagated by distinguishing properties of pixels on both sides of the edge, and tracing a path between pixels with these two sets of properties.

Some algorithms find image regions directly by looking for neighboring regions of pixels with shared properties, rather than by first finding edges. Pixel properties may simply be defined by conditions on red, green and blue pixel intensities. Some image regions do not consist of pixels with similar colors, but are texture regions whose pixels obey some spatial statistics or are a spatial deformation of a repeating pattern (e.g., the regular pattern of a plaid shirt deformed by following the curves of a human body).

Edges and regions are the input to numerous other higher-level vision processes. These processes find the shapes of 3-D objects, track moving 2-D regions or 3-D objects over time sequences of images, and identify objects. Shape is deduced by trying to fit a 3-D shape to the equations that govern how shapes and intensities in 2-D images are determined by shapes of 3-D objects. Motion is tracked by identifying regions and shapes in consecutive images in a way corresponding to steady motion. Objects are identified by matching features of 3-D shapes to a knowledge-base of objects. These higher-level processes interact with each other and with the low-level process of identifying regions. For example a single moving region may appear as two regions in the images for certain time steps and as a single region in the images for other time steps. The motion tracking process can inform the lower-level region identification process that it is always just one region.

Each vision process is trying to invert some part of the physical image formation process, by hypothesizing some physical model that would generate the observed image. For each vision process there will be some error between the image its hypothesized model would generate and the actual image seen by the computer. An individual vision process tries to minimize its error. Some vision systems integrate their different vision processes by a mathematical approach that searches for the minimum total error over all processes.

Vision systems may also integrate vision processes by allowing control to flow up and down processing levels.[19] That is, a low-level process may identify a region, a process at the next level may deduce its 3-D shape from variations in its intensity, and a high-level process may identify the 3-D shape as the front of an automobile. This process may then make a guess about the 3-D shape of the rest of the automobile and invoke the shape deduction process to verify this guess. This in turn may make a guess about 2-D image regions, which can be verified by the region finding process.

Active vision systems integrate vision processes with processes that control camera location and focus. There are inevitable ambiguities in the vision processes that can be resolved by active vision systems that take a look from a different angle, just as a person or animal may move their head to get a different view at a puzzling object. Active vision can greatly reduce the computational requirements for vision processes.

Vision is a very difficult problem. Thus machine learning and neural networks are essential for effective vision systems. In some systems, each pixel is an input to a neural network, with object identifications or system actions as outputs. Such systems completely dispense with the traditional mathematical algorithms of vision processes. Other systems use machine learning to adjust various parameters of mathematical algorithms. Some systems use mathematical algorithms to find basic image features, which are then inputs to a neural network for identifying objects.

An honest assessment is that current AI vision systems are nowhere near human capabilities. You can make your own assessment by sending images to an experimental system at Carnegie Mellon University for detecting faces in images.[20] However, artificial vision systems have one advantage to help them compete with human vision: they do not have the physical restrictions of human eyes. They may combine images from many cameras at different angles, they may look at images using many wavelength bands (not just the red, green and blue bands used by human eyes), and they can employ range sensors such as radar for directly seeing 3-D shape. There are a wide variety of commercial vision applications,[21] demonstrating the utility of the current state of the art in AI vision systems.

ROBOTS

Many scientists think that robots are the proper context for AI research, because robots require AI systems that deal with the realities of physical bodies sensing, moving and acting in the world. This is closely related to the objection of Hubert Dreyfus that AI systems not rooted in physical bodies and lower-level brain functions are doomed to failure.

Robots usually include knowledge in the form of mathematical models of their world and the way their actions cause them to move in and change their world. They usually have cameras for visually sensing the world. Robots have goals. For example, the goals of the soccer-playing robots discussed in a pervious section are to get the ball into their opponent's net but not to let it into their own net. Robots often include machine learning. Learning is nearly essential for effective vision, and for executing sequences of actions in order to achieve goals. Some robots learn mathematical models of their worlds based on their vision and movements.

There is a practical need for robots in the form of unmanned vehicles that can travel on their own to places too dangerous or difficult for humans to go, and where distance or the environment make remote human control impractical. Robots are used in deep oceans, deep space, battlefields, nuclear reactors and environmental clean up sites. The U.S. Defense Advanced Research Projects Agency has supported a long term research program to develop autonomous vehicles.

Carnegie Mellon's ALVINN (Autonomous Land Vehicle in a Neural Network) is an interesting example of a robot that uses a pure artificial neural network system for driving on roads based on input from a low resolution (i.e., 30×32 pixel) camera.[22] The system steers a car and its goal is to keep the car in its lane driving on a highway. This system learns by "watching" a human drive. During the learning phase its inputs are its camera watching the road, and a sensor on the steering wheel being worked by the human driver. The system has learned to drive on a variety of kinds of roads. It has successfully driven at 70 mph over 90 miles of public highway (presumably at a time when the highway was closed to the public).

MACHINE LEARNING

Learning is fundamental to the way human brains work and essential to developing intelligent machines. The deepest and most useful result of AI research has been the development of machine learning techniques.[23]
Machine learning techniques can be divided into several categories:

1. Symbolic. These apply induction to learn new rules for predicate calculus knowledge-bases.
2. Statistical. These apply Bayesian or other statistical techniques to learn rules, numerical algorithm parameters or other forms of knowledge.
3. Reinforcement. These may be symbolic or statistical techniques, but are applied to the general problem of learning how to behave in a world with rewards and punishments.
4. Connectionist. These alter connections in artificial neural networks.
5. Genetic. These learn algorithms by mimicking sexual reproduction and natural selection in a population of competing algorithms.

Symbolic learning is basically induction, which is learning general rules from specific examples. Recall our knowledge-base of the *is-a* predicate. Imagine that a system can directly collect information only about specific individuals. So it might know directly that:

is-a(Tabby, horse)
is-a(Sting, horse)
is-a(Prudie, horse)
is-a(Gertie, horse)
is-a(Tabby, animal)
is-a(Sting, animal)
is-a(Prudie, animal)
is-a(Gertie, animal)
is-a(Sarah, dog)

is-a(Maddie, dog)
is-a(Susie, dog)
is-a(Sarah, animal)
is-a(Maddie, animal)
is-a(Susie, animal)

To learn, the system could look for y and z such that:

$$(\forall x)(is\text{-}a(x, y) \Rightarrow is\text{-}a(x, z))$$

Whenever this is true, the system could induce that *is-a*(y, z). Looking through the information it had collected, it would find that this is true for $y =$ horse and $z =$ animal, and for $y =$ dog and $z =$ animal. But note that it is not true for $y =$ animal and $z =$ dog, since *is-a*(Tabby, animal) does not imply *is-a*(Tabby, dog). Thus the symbolic learning system could induce:

is-a(horse, animal)
is-a(dog, animal)

These rules could then be used to make deductions about new information. So if the system later collected the information *is-a*(Red, horse) it could use the learned rule to deduce that *is-a*(Red, animal).

Statistical techniques can be applied to learn a wide variety of forms of knowledge. For example, a chess playing system may adjust the relative numerical values of different pieces based on experience playing many games. That is, two versions of the system that differ only in the values they assign to pieces can play a long tournament. The judgement that the winner of the tournament has more accurate piece values is statistical.

Statistics are more explicit in systems that learn to classify data. For example, a classical AI problem is to construct a system for filtering text articles for a reader. The system's input is a text article and its output is a binary classification of the article as either interesting or uninteresting to the reader. The system is given a large training set of articles and their classifications by the reader. The statistical approach is to define a number of attributes of the articles. These attributes may be either numerical or non-numerical. Examples include the length of the article, the average length of words in the article, the number of times the word "horse" occurs in the article, the name of the article's author and so on. These attributes define a multi-dimensional space of articles. Points in this space are defined by vectors of attribute values. For example, we may use the 4-dimensional space:

(*article_length, average_word_length,*
number_of_horse_occurences, author_name)

A training set defines a sample distribution of interesting and uninteresting articles over this attribute space. The system will use these to estimate the true distributions of interesting and uninteresting articles over the attribute space, and classify a new article according to whether the interesting or uninteresting distribution has the largest probability for the attribute vector of the new article. This can be estimated using Bayes theorem. Here $P(x)$ denotes the probability that x is true, and $P(x \mid y)$ denotes the probability that x is true given that y is true. Bayes theorem states:

$$P(x \mid y) = P(y \mid x) \times P(x) / P(y)$$

Given a new article, let a be its attribute vector, let I be the assertion that the article is interesting, and let U be the assertion that the article is uninteresting. Then by Bayes theorem:

$$P(I \mid a) = P(a \mid I) \times P(I) / P(a)$$
$$P(U \mid a) = P(a \mid U) \times P(U) / P(a)$$

The system can decide whether the reader will find the article interesting based on which of these is larger. For that, we can remove the common factor $P(a)$. Thus the system needs to compare $P(a \mid I) \times P(I)$ and $P(a \mid U) \times P(U)$. The system can estimate $P(I)$ by the proportion of articles in the training set that were interesting, and $P(U)$ by the proportion that were uninteresting. It can estimate $P(a \mid I)$ by the proportion of interesting training set articles in the neighborhood of a in attribute space, and similarly estimate $P(a \mid U)$. Note these proportions may be difficult to estimate, if attribute space has many dimensions, if the training set is small, and if there are non-numerical attributes such as *author_name*. These problems can be addressed by assuming that the probabilities of different attributes are independent in the interesting and uninteresting distribution. This is not always a valid assumption, but it does make the computation much easier. Under this assumption:

$$P(a \mid I) = P(a_1 \mid I) \times P(a_2 \mid I) \times ... \times P(a_n \mid I)$$

where a_1, a_2, ..., a_n are the individual attributes in a. Proportions of interesting articles are much easier to estimate in each one-dimensional attribute space. A number of practical text classifiers have been designed in just this way.[24]

Classification problems are also addressed by the technique of building *decision trees*. A decision tree defines a sequence of questions about attributes, with branches depending on answers leading to other questions and finally classifications. For example, text articles might be classified by:

> *horse_occurences* > 0
> > *interesting*
>
> *horse_occurences* = 0
> > *author_name* = Groucho Marx
> > > *interesting*
> >
> > *author_name* = Karl Marx
> > > *uninteresting*
> >
> > *author_name* = other
> > > *article_length* > 5000 words
> > > > *uninteresting*
> > >
> > > *article_length* <= 5000 words
> > > > *average_word_length* > 6
> > > > > *uninteresting*
> > > >
> > > > *average_word_length* <= 6
> > > > > *interesting*

Given a training set, statistical techniques can used to produce the smallest decision tree that is consistent with all the samples in the training set.[25]

Machine learning is important in most practical applications of AI, and will be essential in building machines with real intelligence. The next three sections discuss learning techniques modeled after the way human and animal minds learn, and the way that natural selection "learns" to evolve effective species.

REINFORCEMENT LEARNING

Reinforcement learning occurs in the general situation when an autonomous agent (e.g., a person, animal or robot) acts in a world with rewards and punishments, seeking to maximize rewards and minimize punishments. This is a very complex problem, since:

1. The agent may not start with a complete description of the world, but may need to interact with it (i.e., explore) in order to form a model of how the world works. For example, people and animals are born knowing little, and must learn about the world as they grow.
2. The agent may not be able to observe the full state of the world at any moment.
3. The rewards and punishments may be delayed. If a person or animal eats bad food, it may take hours until they become ill. This is known as the temporal credit assignment problem, and is a central problem for learning intelligent behavior in a complex world.
4. Reinforcement learning never stops.

One approach to reinforcement learning is called the Q-learning algorithm.[26] The agent constructs a model of the world, including a set of world states S, a function $n(s, a)$ that defines the next state in S when agent applies action a to state $s \in S$, and an immediate reward function $r(s, a)$ that assigns a numerical immediate reward value (negative values for punishment) for applying action a to state $s \in S$. The immediate reward is just the reward that occurs immediately on executing action a from state s. We assume that the agent can observe the value of the immediate reward. The problem is to estimate the long term reward $Q(s, a)$ that results from taking action a from state s. The goal of reinforcement learning is a function $B(s)$ that defines the optimal action a to take in state s.

The Q-learning algorithm recognizes that $Q(s, a)$ satisfies the following equation:

$$Q(s, a) = r(s, a) + c \times \max\{Q(n(s, a), a') \mid a'\}$$

Here the maximum ranges over all actions a' that can be applied to state $n(s, a)$, and c is a non-negative constant less than 1.0 that defines the discount for future reward. The agent starts with an estimate of the values of $Q(s, a)$, which may just be $r(s, a)$, and repeatedly explores the state space S

by applying a sequence of actions. When it takes an action a at state s, the agent updates its estimate of $Q(s, a)$ by:

$$r(s, a) + c \times \max\{Q(n(s, a), a') \mid a'\}$$

This algorithm gradually propagates reward information back through sequences of states. Under certain circumstances, it can be proved that eventually the estimates of the values of the Q function will converge on their real values. Given an estimate of the Q function, the optimal behavior function B can be computed as:

$$B(s) = a \text{ that maximizes } Q(s, a)$$

There are many variations on this basic algorithm, for different exploration strategies, for only partially knowable states, and for non-deterministic functions for next state and immediate reward. The Q-learning algorithm is widely used in robots.

Another approach to reinforcement learning is the temporal difference (TD) algorithm. The TD algorithm can be presented as a generalization of the Q-learning algorithm that looks ahead by multiple steps when updating the estimate for the Q function. However, its development was motivated by a desire to model observed learning behavior in animals. In this perspective the TD algorithm can be presented as a specific design for a network of simulated neurons. Neural network simulations are described in the next section, but a description of the TD algorithm is deferred to the next chapter on neuroscience because of its importance for understanding how humans and animals learn.

ARTIFICIAL NEURAL NETWORKS

Artificial neurons are an abstract mathematical model of human and animal neurons. They accept numerical inputs from other neurons, and produce a single numerical output. The neuron computes a weighted sum of the values of its inputs, and applies a non-linear function to this sum to compute the value of its output. That is, given a set of input values $x_1, x_2, \ldots, x_n$, the output is:

$$f(w_1 \times x_1 + w_2 \times x_2 + \ldots w_n \times x_n)$$

Here the w_i are the weights and $f(x)$ is a non-linear function of x. The earliest neural network experiments used a simple threshold function:

$f(x) = -1$ for $x < t$
$f(x) = 1$ for $x >= t$

This threshold function is meant to model the notion of discrete neuron firing. However, experiments with artificial neural networks have shown that they are more effective with continuous functions of the form:

$$f(x) = 1 / (1 + e^{-x})$$

This function is approximately 1 for large positive values of x, approximately 0 for large negative values of x, and has a smooth transition near $x = 0$.

Artificial neurons are usually connected together in layers. A common approach is to use a layer that consists simply of raw inputs, connected to a hidden layer, then an output layer. For example, the ALVINN autonomous vehicle described in the section on robots has an array of 30×32 inputs for the pixel intensities coming from its camera. Each of these is an input to each of 5 neurons in the hidden layer. The output of each hidden layer neuron connects to a layer of 30 output neurons. The output neurons represent 30 positions of the vehicle steering wheel, which turns in the direction of the output neuron with maximum output value.

Artificial neural networks learn by adjusting the weights of their neurons, based on training exercises that tell the network what its outputs should be for given inputs. For the neurons in the output layer, the weight for each input is negatively adjusted by an amount proportional to the output error and the input value. Specifically, if o is the actual output, d is the correct output and x_i is the i-th input, then the change to the i-th weight is:

$$\Delta w_i = -c \times (d - o) \times o \times (1 - o) \times x_i$$

Here c is a constant that controls the speed of learning. For neurons in hidden layers, the corresponding formula is:

$$\Delta w_i = -c \times (\Sigma_j(-delta_j \times w'_j)) \times o \times (1 - o) \times x_i$$

Here the sum Σ_j is taken over all neurons j whose inputs include the output of this neuron, w'_j is the weight of this neuron as an input to neuron j, and delta $_j$ is the $(d - o) \times o \times (1 - o)$ value from neuron j. This technique for adjusting weights in multi-layer neural networks is called *backpropagation* (because learning propagates backward through the neuron layers).

We can imagine the set of all neuron input weights in a network as defining the coordinates of a multi-dimensional space. For any point in this space we can assign numerical error values that a network with such weights would make in its outputs. The mathematical details of backpropagation can be derived from a desire to adjust all weights in the direction in this space toward zero errors.[27] The constant c governs how fast weights move along that direction.

Most practical applications of artificial neural networks use multi-layer feed-forward networks with backpropagation learning, as described here. Feed-forward means that outputs from one layer do not feed back to the inputs of previous layers. However, human and animal brains certainly have such feedback and some AI researchers are experimenting with networks with feedback. However, so far such networks are less effective for practical applications.

Human and animal brains use a variety of learning techniques, including Hebbian learning.[28] This strengthens the weight of a connection between two neurons when they both fire at approximately the same time. Clearly this would be useful for remembering the connections among simultaneous events, and has been used for experiments with associative memory systems.

The human brain has many innate general abilities, which are made specific via changes in neural connections. For example, brains seem to have innate abilities for recognizing human faces and for language, but these must learn specific faces and specific languages. Intelligent machines will be similarly programmed for general tasks, but will also need to learn the specific details of those tasks via artificial neural networks. Artificial neural networks have demonstrated impressive success in a wide variety of applications, and will be essential for developing truly intelligent machines.

GENETIC PROGRAMMING

Genetic techniques learn algorithms by mimicking sexual reproduction and natural selection in a population of competing algorithms.

The field divides into two variants: genetic algorithms represented by sequences of bits, and genetic programs represented by tree structured programs. However, both variants require the following elements:

1. A problem to be solved.
2. A population of programs (or algorithms).
3. A measure of the fitness of any program for the problem to be solved, used to select programs for reproduction.
4. A means of combining two programs to produce offspring for the next generation.
5. A termination condition, to determine when the problem is adequately solved.

Programs inside computers are represented by bits, but combining two programs by splicing together some bits from one and other bits from the second will generally result in programs that are illegal or don't run. So the first issue is to find representations for algorithms and programs, and ways of combining them, so that the results of combination are at least viable. Research with genetic programming has shown that finding the right representation is also important for the natural selection process to produce adequate programs in a number of generations that is feasible to compute.

One interesting example uses genetic programming to learn a program for stacking blocks.[29] This models a set of blocks and at any instant some subset of blocks form a stack. The problem uses a very small programming language with operations specialized for manipulating the stack of blocks. Example operations include:

(stack x)	move block x to top of stack,
	returns true if this is possible, false otherwise
(unstack x)	remove top block if block x is in stack,
	returns true if this is possible, false otherwise
top	return name of top block on stack,
	or false if stack is empty
(equal x y)	return true if x = y, false otherwise
(do x y)	do expression x until expression y = true

Using this language, a program to remove all the blocks from the stack would be:

(do (unstack top) (not top))

This simple program unstacks the top block until there is no top block. Two programs "reproduce" by replacing expressions (i.e., text inside matching parentheses) in one program by expressions from the second program.

The fitness of programs is measured by running them on a large number of initial block and stack configurations, and calculating what proportion of these are correctly stacked. The probability of a program engaging in reproduction with other programs is proportional to its fitness. Also, the best programs from one generation survive into the next generation (i.e., they join their offspring in the next round of competition).

In this experiment there was an initial population of 300 randomly generated programs. Fitness was judged against a set of 166 initial block and stack configurations. After only 10 generations the population included a program that worked correctly on all 166 test cases.

Genetic programming shows promise, but it has not yet produced the sort of practical successes that machine learning and artificial neural networks have.

CONCLUSION

Artificial intelligence research has produced a variety of successes and even commercial products. However, it has not produced anything like human intelligence. Learning is of fundamental importance to intelligence, and the most valuable results of AI research are the lessons we have learned about how machines can learn.

NOTES

1. Brooks, 1999.
2. Samuels, 1959.
3. Goodman and Keene, 1997. http://www.research.ibm.com/deepblue/.
4. Yu et al, 1979.
   http://www.computing.surrey.ac.uk/research/ai/PROFILE/mycin.html.
5. http://www.askjeeves.com/
6. Unger, 1988.
7. http://www.cyc.com/

8. http://www.cs.utexas.edu/users/mfkb/bkb.html
9. Winograd, 1972.
10. Kitano, 1997. http://www.RoboCup.org/
11. Robinson, 1965.
12. Rickel and Porter, 1997.
    http://www.cs.utexas.edu/users/mfkb/tripel.html
13. Lester and Porter, 1997.
    http://www.cs.utexas.edu/users/mfkb/knight.html
14. Katz, 1997. http://www.ai.mit.edu/projects/infolab/
15. http://www.babelfish.org/
16. http://www-3.ibm.com/software/speech/
17. http://www.speech.cs.cmu.edu/sphinx/
18. Ballard and Brown, 1982.
19. Binford, Levitt and Mann, 1989.
20. http://vasc.ri.cmu.edu/cgi-bin/demos/findface.cgi
21. http://www.cs.ubc.ca/spider/lowe/vision.html
22. Pomerleau, 1993.
23. Mitchell, 1997.
24. Lang, 1995.
25. Mitchell, 1997.
26. Watkins, 1989.
27. Luger, 1997.
28. Hebb, 1949.
29. Koza, 1992.

Chapter 6

NEUROSCIENCE

The human brain is our only example of an implementation of intelligence, so neuroscience has much to teach us about how real intelligence works. However, before getting into any details of scientific understanding of brain functions, we note a couple very general characteristics of brains. They are non-linear systems with many dimensions, which means that they are chaotic and exhibit the *butterfly effect*. That is, very small events cause large events. The name butterfly effect is taken from meteorology, in which a butterfly flapping its wings can cause a hurricane to develop months later.[1] In addition to being non-linear, brains are open systems. They are open to all sorts of sensory input, as well as cosmic rays and the random thermal noise of their molecules. In the brain, non-linearity and openness mean that a cosmic ray colliding with molecules in a person's brain can cause that person to decide to go to medical school. And that cosmic ray may have been created in another galaxy billions of years ago. Thus any debate over whether brains are deterministic or non-deterministic is academic. Whether a person goes to medical school or becomes a bond trader can depend on the collision of atomic nuclei long, long ago in a galaxy far, far away. In any useful sense, brains are non-deterministic.

Whether this practical non-determinism is the same thing as free will depends on how you define it. Some define free will by saying that minds make decisions without any cause. I think that minds are explained by physical brains and all physical events have physical causes, so we do not have free will in the sense of thoughts without causes. On the other hand, it is impossible to predict physical events in non-linear systems like the brain, so we do have free will in the sense that our thoughts cannot be predicted. There is no practical reason to try to settle this; it is largely a matter of how you define free will. The concept of free will arose from the lack of perception of low-level events in our brains that cause our perceptible thoughts. What matters is that people can make decisions independently of what others tell them and in many cases can escape the general pattern of their families and cultures. Some individuals even make creative contributions to knowledge and culture. This is a good practical substitute for whatever free will means.

The human brain contains about 100 billion neurons, connected together at about 100 trillion synapses to form a network. And each neuron is complex: I have heard a brain researcher claim that each neuron has the complexity of a Pentium computer chip. Thus brains are much more complex than current computers. The neural network is irregular in its details, but also exhibits regularities. For example, sensory neurons from retinas (in the eyes) and skin are mapped onto areas of the brain in ways that preserve spatial continuity. That is, points on the skin or retina that are close together are mapped to points in the brain that are close together. Furthermore, specific brain areas are responsible for specific behaviors. A part of the brain called Broca's area plays a key role in generating speech, while another called Wernicke's area plays a key role in comprehending speech.[2] The associations between brain areas and behaviors are established by observing behavioral changes in people with injuries to specific brain areas, and electrical activity in specific brain areas during specific behaviors. There are currently many scientists studying brain function, and they are generating detailed knowledge of the behavioral roles of various brain areas (however, mental behaviors involve neurons distributed over many different brain areas, and there are differences in brain-mind correlations between different people).

As previously mentioned, the existence of so many correlations between brain functions and mental behaviors tells me that ultimately brains must explain minds. For if minds ultimately have a non-physical explanation, then these correlations would be absurd coincidences.

Neurons have periodic events called *firings*, which are sudden changes in their electrical potential and chemistry. When one neuron fires, it makes small changes in the other neurons that it is connected to. These changes either increase (excite) or decrease (inhibit) the likelihood of these other neurons firing (but some neurons do not fire suddenly but rather interact with other neurons via gradual electrical and chemical changes). These causal relations between the firings of connected neurons give rise to incredibly complex patterns of neuron firings in our brains.

Light coming into our eyes and sound coming into our ears causes neurons to fire. And firing neurons cause other neurons to fire and ultimately cause muscles to contract and our bodies to move. This can be as simple as jerking our arm back from a hot stove, or as complex as subtle body language when we see someone we are attracted to. In the case of the hot stove reflex, the chain of neuron firings is so direct it doesn't even pass through our brain. But in the case of being attracted to someone, the neuron firings in our optic nerve trigger a sequence of neuron firings in many different brain areas to: recognize the image as a person, associate features of that person with characteristics that attract us, possibly invoke pleasant memories, activate behaviors to attract others, and finally control our body motions.

One very important characteristic of a human brain is that its neuron network is constantly changing. These are changes in network topology, so that neurons make and break connections with other neurons. There are also changes in the strengths of connections, increasing or decreasing the influence of neurons on other neuron firings. These changes are the way memories are stored and new behaviors are learned.

The enormous complexity of neuron interconnections and their relative strengths encode everything a person remembers and everything they know how to do. But the encoding of memories and behaviors in neuron connections cannot be understood or decoded like the bits in a computer memory. A single bit of information, such as the gender of an acquaintance, is diffused through many neurons rather than encoded in a single neuron connection. In fact, that bit of information may be dynamically recomputed when needed, by recalling images and behaviors of the acquaintance and answering the gender question as if seeing the person for the first time.

Furthermore, when our brains reconstruct images and behaviors of others they may be based on only a few details, with the rest filled in according to our general worldview. Our brains have this tendency to fill in details even when we are actually observing something directly. That is, we jump to conclusions. This is illustrated by optical illusions in which we can see the same image in two different ways. The tendency to fill in details is also responsible for cases where people can swear to memories that are simply wrong.

Much of what we know and the way we reason is ultimately stored as memories of our sensations and actions. That is, even abstract thoughts such as mathematics are played in our brains as experiences with the way that diagrams or equations look, or the way that spoken formulas sound. Our reasoning is a set of associations for how we can manipulate and combine these sensual experiences. Our mental processes are rooted in our physical experiences. This is reflected by brain research that finds that merely thinking about a sight or sound triggers the same neurons that are triggered by the actual sight or sound.

BRAIN STRUCTURE

Brains are not random networks of neurons. Rather, they are divided into numerous specialized regions with identified roles in mental behavior, and connected to each other in specific ways. Different regions consist of different kinds of neurons, which are distinguished by their shapes, connectivities, and the specific chemical mechanisms of their firings.[3] There is less detailed understanding of the specific behaviors of specific neuron types, with the best understanding for neurons most closely related to sensory input and motor output.

The human brain is structured in layers.[4] The lowest layers, the hindbrain and midbrain, evolved earliest and control basic functions necessary for survival like digestion, respiration, blood circulation and temperature maintenance. These layers also control coordination of movement and participate in higher mental functions in ways that are not fully understood. The highest layer is the forebrain, which includes the thalamus, hypothalamus and cortex. This layer is the seat of higher mental functions such as language and problem solving but also plays an important role in regulating basic survival functions.

Another classification, more clearly aligned with evolutionary stages, divides the brain into a reptilian brainstem core, the limbic system (old mammalian brain) and the neocortex (new mammalian brain). The reptilian core implements innate behaviors necessary for survival, whereas the limbic system supports basic learning of behaviors. The limbic system includes the olfactory bulb, responsible for the sense of smell, the hippocampus, important for memory, and the amygdala, which plays an important role in emotions.

The neocortex is a sheet of six distinct layers of neurons, about 1/8 inch thick, on the outer surface of the brain. All the folds of the brain surface serve to increase the area of the neocortex. Human brains are distinguished from more primitive animal brains by the degree of folding and surface area of the neocortex.

Nerves from areas of the body are connected to areas of the neocortex in topological maps. That is, adjacent areas of the body connect to adjacent areas of the neocortex. Numerous topological maps connect nerves from the eyes to different regions of the neocortex. These different regions are responsible for different aspects of visual processing, such as shape, color and motion. There are connections among these various visual regions of the neocortex, for the integration of higher level visual understanding. In fact, all higher level mental functions map to various regions of the neocortex.

There are numerous connections between various regions of the brain.[5] There are lateral connections among different areas of the neocortex, as well as loops from the neocortex to the thalamus and then back out to the neocortex. These appear to be involved in associations between various sensations and memories. Different types of memories, such as events, values and behaviors, are stored in different brain areas. When we see or think of a person, place, thing or event, these connections among neocortex and thalamus neurons trigger numerous related memories.

Another set of connections go from the neocortex to the cerebellum in the hindbrain that provides fine control over motion, and back to the neocortex. A related set of connections go to the basal ganglia, whose role is not completely understood but relates to motion planning and learning. And another set of connections loops from the neocortex to the hippocampus, which is involved in generating long-term memory from short-term memory.

The brain includes a number of very small nuclei in the lower brain and hypothalamus that control innate values or emotions. These contain only thousands of neurons, but connections fan out from these nuclei over the entire brain. They release various chemicals called neuromodulators that alter the activity and plasticity (i.e., ability to change) of neurons. Some of these, such as noradrenaline, may prepare the brain for quick action. Others, such as dopamine, enable quick changes in connections between neurons to positively or negatively reinforce behaviors. These nuclei and their connections to the rest of the brain provide a physical implementation of the mechanisms of emotion and learning. We can associate the emotions of fear or anger with the heightened brain activity necessary for fleeing or fighting. And we can associate curiosity, frustration and in fact any emotion with the activity of learning.

Our brains are more plastic when we are young, to accommodate a greater rate of learning. But brains can make significant changes even in adults. For example, when brain regions are injured, the brain compensates by forming new connections in other regions. Injuries in visual areas can cause visual connections to grow in brain areas previously used for sensing touch.

As Steven Pinker writes in *How the Mind Works*, far from being abstract learning machines, animal and human brains include many innate, complex computational algorithms.[6] For example, baby birds in their nests watch the night sky to learn the patterns of star constellations and how they move. They learn the location of north (or south if they live in Australia) as the point in the sky that stars rotate around, and they learn which constellations are near the north point. Then as adults they use this knowledge of stars to guide their migrations. This amazingly complex behavior is innate in their brains. Remember that the next time someone says "bird brain."

Humans do not migrate as birds do and modern humans sleep under roofs that block the night sky, so we have no such behavior. But human brains seem to have innate image processing algorithms that enable them to learn to recognize human faces and interpret their emotions, and to learn to construct 3-D models of the world around them. In fact, human brains include thousands of complex innate computational algorithms. Innate human algorithms include learning just as birds' innate star algorithms include learning. The human brain can be thought of as a collection of many specialized information processing organs intimately connected together, each capable of learning. The artificial intelligence pioneer Marvin Minsky says he bets the human brain is a kludge (computer programmer slang for an inelegant design).

Finally, the brain is a network of neurons connected by synapses, and we can analyze its mathematical structure as a network. Biologists have studied brain sizes of various mammals over four orders of magnitude (from mice to whales) and found consistent patterns in the way network structure varies as a function of brain volume.[7] Gray matter consists of neuron cell bodies in the neocortex, and the synapses connecting them. White matter consists of axons, which are long extensions of neocortex neurons that enable them to connect to distant neurons. Connections between neurons that are physically near each other exist entirely in the gray matter, whereas non-local connections require axons in the white matter. Studies of mammal brains show that the volume of white matter is roughly proportional to the volume of the gray matter raised to the 1.3 power (i.e., a brain with 2 times as much gray matter will have about 2.46 times as much white matter). Thus as the number of neocortex neurons increases, the number of non-local connections per neuron increases. The diameter of a network like the brain can be defined as the average number of synapses that have to be traveled to get from one neuron to another (e.g., to fly from Madison to New York I have to travel two legs: one from Madison to Milwaukee and a second from Milwaukee to New York). If mammal brains all had the same number of connections per neuron then larger brains would have larger diameters. However, because the connection density increases with brain size, the diameter of brain networks seems to be roughly constant at about 2.6. Some scientists speculate that this diameter is the optimal trade-off between efficiency of learning and avoiding feedback oscillation.

VISION

Half of the neurons in our brains process information from our eyes. There are roughly 30 different areas of the neocortex devoted to different visual functions. These are two-dimensional areas on the surface of the brain, and about half of them are topological mappings from the two-dimensional retinas. They are roughly organized in a hierarchy from low level functions, such as recognizing lines, to higher level functions, such as recognizing faces. They are divided into two major pathways: one concerned with recognizing objects and the other concerned with recognizing where the objects are and their spatial relationships. In addition to the topological maps of retinas, there are also egocentric maps concerned with location relative to the viewer (compensating for head and eye position), and allocentric maps that are independent of the viewer (e.g., a map of roads in the viewer's town).

There are fascinating patterns to the arrangement of neurons in the low-level vision area known as V1 (which is part of the neocortex near the back of the brain). Signals from the retina come through the thalamus to layer 4 of area V1 (recall the neocortex has six layers). There is a topological mapping of the 2-D retina to the 2-D area V1, and cells in layer 4 respond to small spots of light at the point of the retina that they map to. But cells in other layers respond to lines of light across the retina. Area V1 is divided into bands of cells responding to different orientations of these lines. These are further organized into larger bands alternating response to left and right eyes. And there are blobs of cells embedded in this band structure that respond to colors. Neurons from area V1 send signals to other visual areas that respond to higher-level visual stimuli, such as motion and form, and then to the major pathways for recognizing objects and for spatially locating objects.

In some remarkable experiments, the brains of young ferrets have been rewired so that visual signals from their retinas do not connect to their visual cortex area V1, but instead connect to their auditory cortex.[8] The same banded structure of cells specialized for different line direction orientations that normally develop in the visual cortex develop in the auditory cortex. This suggests that these structures are not encoded in ferret genes, but rather are learned in response to the structure of visual stimuli. This suggests that very little about brain behavior is innate. Rather it is learned through interaction with the world.

As neuronal signals pass through the thalamus on their way to the neocortex, they encounter cells that can mask off parts of the 2-D scene falling on the retinas. These may play a role in focusing visual attention on one object at a time.

Given that different brain areas process color, form and motion, a central problem is to understand how the brain binds these properties together in our mental pictures of objects. This is called the *binding problem* and will be discussed in more detail in the section on consciousness.

There are other neural pathways from the retina that bypass the neocortex altogether, going to areas in the midbrain. These probably play a role in blindsight, which is the ability of some patients with injuries to visual areas of their neocortexes to behave with knowledge of object locations without being conscious of that knowledge.

LANGUAGE

There are strong similarities between all human languages, which have led to theories of innate language behavior in the human brain. But there is a debate over how specific to language those behaviors are and whether their evolution was selected only because of the advantages of language. Languages all have vocabularies of thousands of words and they all use the same basic parts of speech. When people from different cultures mix the adults invent a crude pidgin language. Their children invent a more complex creole from the pidgin, which always uses a subject–verb–object sentence order. Rather than reflecting language-specific behavior, these similarities may reflect the structure of the way human brains perceive and plan actions. And the similarity of vocabulary size may simply reflect the capacity of human brains.

There are specific brain areas associated with language. In the left hemisphere, Broca's area controls speech generation and Wernicke's area controls speech comprehension. However, speech processing can develop in other areas in children who have brain injuries in their left hemispheres. Furthermore, there are no observed cases of genetic mutations that degrade language behavior but leave all other behaviors intact. Thus it is likely that language development exploits more general brain behaviors for recognizing objects and reasoning about relationships between objects.

Even if language behavior is derived from more general behavior, the improved social teamwork enabled by language was probably a factor in the natural selection of individuals with the more general behavior. In fact, human language has replaced genetics as the primary communications medium for the progress of life on earth. Language provides the communication necessary for groups of humans to work together on complex tasks, and the communication over time for humans to build on the accomplishments of previous generations.

The specific brain areas associated with language behavior communicate with numerous other areas that provide meanings for language. Knowledge of people, animals and tools exists in different brain areas, which become active when language about their subject matters is processed. Speaking about color activates brain areas used to perceive color, and speaking about action activates brain areas that perceive and plan motion.

BRAIN MODELS

The current barrier for brain research is correlating behavior with the actions of neurons. This is a barrier because information is so diffuse in the brain and because we do not yet possess instruments for observing more than a few individual neurons. Neural modeling is necessary to understand behaviors of large numbers of interacting neurons.[9] Experimental simulations of large numbers of interacting neurons can be compared with observed human mental behaviors, as a way to find neural organizations capable of explaining mental behaviors. Some research is using feed-forward networks with backpropagation to implement reinforcement learning, as described in the previous chapter.

One excellent example is a feed-forward neural network model that learned associations between images and words.[10] The model was then tested in two tasks: its ability to recall a word when it is given an image, and its ability to produce an image when given a word. The model accurately simulated a number of characteristics of human learning behavior, including:

1. It was better at producing images from words than recalling words for images.
2. The network exhibited a sudden increase in the number of words (i.e., vocabulary) for which it could accurately perform both tasks.
3. At the time of the sudden increase in its vocabulary, the network exhibited a change in the types of errors it made for both tasks.
4. Most remarkably, the model even predicted a formerly unknown characteristic of human learning behavior that was subsequently confirmed by observing learning behavior of children.

In another experiment, a feed-forward neural network model learned sentence structures with relative clauses.[11] The model could only learn to accurately parse such complex sentence structures if it first had the opportunity to learn simple sentence structures. However, it could simulate this sequence of learning simple then complex sentence structures by starting with limited memory, then increasing memory. This experiment suggests that young children's limited memory may actually help them learn sentence structures.

Neural network models can also simulate the effects of damage and compare these with the behavior of human patients with brain damage. A neural network model of dyslexia exhibited the linkage between visual errors and semantic errors also observed in human patients.[12] The model learned to associate images of printed words with semantic features (e.g., color, use or composition) of the objects named by the words. The key point is that a simulated injury to the network caused errors in both words and semantic features. This was not a feed-forward network, but an *attractor network*. Such networks incorporate feedback among their simulated neurons. This feedback causes reverberations among neurons in the network. New input, such as a sensed image of printed words in this dyslexia experiment, triggers reverberations which eventually settle down (assuming the network is damped) on a new steady state or simple oscillation (the attractor) at which point the network's output is useful. In this experiment, damage to the attractor network was simulated by removing some of its neurons or connections. The success of this experiment suggests it is likely that human brains, with extensive feedback loops among their neurons, use attractor networks. Brain attractor networks were presaged by Karl Pribram's ideas about holographic processes in the brain.[13]

Human and animal brains do not function solely by electrical connections among neurons. Some neurons release chemicals called neuromodulators that change the behavior of other neurons. Models of neuromodulation are essential for accurate modeling of brain functioning.

In addition to modeling human behavior, artificial neural networks are also used for practical applications. A great deal of theory and experimentation has gone into designing artificial neural nets that are effective, and in fact these neural nets have learned to compete with human decision makers in fields like stock market investing and certain types of medical diagnosis. There are even commercial neural net chips.

The practical success of artificial neural networks of only a few thousand artificial neurons demonstrates the power of neural networks to process information, and indicates the plausibility that the network of 100 billion neurons in the human brain can explain human behavior.

LEARNING AND EMOTIONS

In order to survive and reproduce, animals must eat, stay warm, mate, avoid danger and perform a variety of other behaviors. Such behaviors are motivated by values, such as "food is good," "warmth is good," "sex is good" and "getting hurt is bad." These values are what we call emotions.

Some behaviors are innate, so for example even the simplest animals know how to eat. But in complex animals many more behaviors are learned. For example, my cat has learned that the cornfield is a good place to catch and eat mice. In higher animals, emotions motivate behavior primarily by reinforcing the learning of behavior. My cat's behavior of going to the cornfield was positively reinforced by eating more mice. The common dog behavior of chewing on lamp cords is negatively reinforced by painful electrical shocks. The strengthening and weakening of connections in artificial neural networks are models of this reinforcement learning in animals and humans.

Like eating, breathing and blood circulation, learning is an innate behavior of the human brain.[14] In fact, learning and the emotions that define the reinforcing values for learning are essential for intelligence. The Spock character on Star Trek, who is supposed to combine great intelligence with a complete absence of emotions, is impossible. But of course, the point of many Star Trek episodes was that Spock did have emotions, even if he hid them.

While innate human emotions reinforce learning of behaviors, they also reinforce learning of new emotions. Human babies are not born wanting a job promotion, but as they mature they learn that job promotions help in achieving their innate values and thus become values in their own right. Adult humans learn very complex webs of interacting emotional values. In a sense, innate behaviors and innate emotions form the "program" of the human brain, and everything else is learned via interaction with the world.

Most computer programs simply define a series of steps for doing something useful. The behaviors of such programs do not change with experience. They do not learn. They are analogous to simple innate animal behaviors. However, very complex computer problems such as speech recognition (i.e., converting spoken language into printed language) are easiest to solve by writing programs that learn. Typically such programs use artificial neural networks or statistical models, and have training phases during which they receive feedback about their accuracy. They use this feedback to amend the way they make choices. A problem like speech recognition could be solved with a non-learning program, but it would be more difficult to write and less accurate than a learning program. Speech recognition is difficult because the variety of ways that different people speak is not well defined. Speech is a complex reality without a neat mathematical characterization. A learning approach constructs a characterization (usually a messy one rather than a neat one) and can update that characterization as speaking habits evolve (i.e., as new words and expressions come into common use, or if the listener gets to know someone with an unfamiliar accent).

Speech recognition is only a small part of the overall problem of succeeding in society faced by modern humans. This problem is so complex and changeable that a solution without learning is really unimaginable. Thus learning is an essential behavior for intelligent minds. This is analogous to the answer to Searle's Chinese Room argument. In that case, the Chinese Room is theoretically one way to mimic human language behavior but in reality one that cannot be constructed. Similarly, while a non-learning machine might theoretically mimic human behavior, in practice it would be a prohibitively difficult and expensive solution.

The primary learning mechanism of neurons is called Hebbian learning.[15] It is a strengthening of the synapse connection between a firing neuron and a second neuron whose firing helped cause the first to fire, and an associated weakening of synapse connections under other circumstances. In some cases Hebbian learning can implement reinforcement learning. For example, when someone touches a hot stove the negative reinforcement is immediate and comes while the neurons causing the action of touching the stove are still firing. However, in many cases reinforcement occurs later than the behavior to be reinforced. This is the temporal credit assignment problem discussed in the previous chapter, and is the central problem for learning intelligent behaviors in a complex world.

The temporal difference (TD) algorithm introduced in the previous chapter is one approach to solving this problem. The TD algorithm can be expressed as a neural network, and in fact observed behaviors of certain types of neurons in animal brains resemble elements of the TD algorithm so closely that there is no doubt that brains do implement the TD algorithm. We will describe the TD algorithm by a set of neural signals, represented as functions of time t. This description closely follows the observed implementation of the TD algorithm in animal brains. The input signals are:

1. $stimulus(t) = 0$ except $stimulus(ts) = 1$. This is an external stimulus, sometimes called a conditional stimulus, firing at time ts. In an animal experiment it may be the neural reaction to a bell ringing.

2. $reward(t) = 0$ except $reward(tr) = 1$. This is an external reward, sometimes called an unconditional stimulus, firing at time tr. In an animal experiment it may be the neural reaction to food delivery.

Note that $ts < tr$ (and there may be no value for tr if the reward is never delivered). The output signal is:

3. *reinforce(t)*. This is sometimes called a reward prediction and reflects the animal learning to associate the *stimulus* with the future *reward*.

We also define some internal signals:

4. *wm_stimulus(t)*. This is a working memory of the stimulus, firing from the time *ts* of the stimulus until the time *tr* of the reward.
5. *delay_stimulus(d, t)*. This is a delayed version of the stimulus signal, indexed by various delay amounts *d*.

The TD algorithm also defines some weights that reflect the strengths with which certain neural signals influence other neural signals:

stimulus_weight = weight of *stimulus(t)* in causing *reinforce(t)*
delay_weight(d) = weight of *stimulus(t)* in causing *delay_stimulus(d, t)*

The non-input signals represent neurons in the TD algorithm implementation, and the weights represent strengths of certain synapses of these neurons. The non-input signals are computed by the rules:

wm_stimulus(t) $\quad$ = 1 for $ts \le t < tr$
$\qquad\qquad\qquad$ = 0 for $t < ts$ or $tr \le t$

delay_stimulus(d, t) = *delay_weight(d)* × *stimulus(t − d)*

reinforce(t) $\qquad$ = *stimulus_weight* × *wm_stimulus(t)*
$\qquad\qquad\qquad$ + *reward(t)*
$\qquad\qquad\qquad$ − sum{*delay_stimulus(d, t)* | $d > 0$}

The synapse weights are updated by the rules:

if (*reinforce(t)* = 1 AND *delay_stimulus(d, t)* = 1) *delay_weight(d)* = 1

if (*reinforce(t)* = 1 AND *wm_stimulus(t)* = 1) *stimulus_weight* = 1

The weights all start at 0, so on the first trial the *reinforce(t)* signal is identical to the *reward(t)* signal. After training, the weights become 1 and the *reinforce(t)* signal is identical to the *stimulus(t)* signal.

External to the neural network implementing the TD algorithm, we assume that there is some neuron weight by which the *stimulus(ts)* = 1 will trigger the action that causes the external reward to be delivered (e.g., the animal pressing a bar). The *reinforce(ts)* signal learns to deliver its pulse at the same time the *stimulus(t)* signal does, and we assume that it increases the weight of the stimulus triggering the successful action just as it strengthens the TD algorithm's internal weights. Note that the input *stimulus(t)* signal may be gated by the appropriate action, so that reinforcement is specific to that action.

In all these cases of neuron weight strengthening, the change depends on the neuron input being active at the same time that the *reinforce(t)* signal is active. This is how Hebbian learning works in animal neurons. In experiments with monkeys, the *reinforce(t)* signal closely resembles the behavior of neurons in the substantia nigra pars compacta (SNc) that generate the neurotransmitter dopamine, and dopamine does indeed enhance Hebbian strengthening of synapse weights.[16] Furthermore, the *delay_stimulus(d, t)* signals resemble the behavior of neurons in the striosomes, and the *wm_stimulus(t)* signal resembles behaviors of other neurons. The entire TD algorithm implementation is centered on the basal ganglia and associated midbrain structures.

This simple TD algorithm implementation in the basal ganglia can learn a sequence of actions, by working backward through the sequence and using the *reinforce(t)* signal for one action as the *reward(t)* signal for the previous action. To work effectively this implementation requires short and predictable delays between stimuli and rewards. This is appropriate for learning sensorimotor skills and there is evidence that this basal ganglia TD algorithm implementation is involved in learning motion skills. However, the world presents humans with many situations where effective behaviors can only be learned by solving the temporal credit assignment problem where time delays are not short and predictable. The theme of the next section is that consciousness is the brain's solution to this problem.

CONSCIOUSNESS

Consciousness is the ultimate problem of neuroscience. Scientific evidence of consciousness depends on verbal reports from subjects, which are more difficult to measure than physical quantities like electrical and chemical activity in neurons. This makes it difficult to precisely define what consciousness is, and makes it difficult to decide whether animals are conscious. The problem of consciousness has been fundamental in the development of human philosophy. Plato speculated that the external physical world is just an illusion, and that only our internal mental world is real. The dualism of René Descartes asserts that the gap between subjective and objective cannot be bridged. Understanding human consciousness in terms of the physical behavior of neurons will bridge this gap. Many people cannot accept that this gap can be bridged, and hence they cannot accept that machines can ever be conscious. They cannot imagine that a machine can ever have a sense of the blueness of blue objects in the way that humans do.

For many years neuroscientists rejected attempts to explain consciousness, because of the philosophical danger that the question was too subjective to have meaning. However, in 1984 Francis Crick began an earnest attempt to explain consciousness.[17] His Nobel Prize for the discovery of DNA commanded such respect that other scientists had to take his attempt seriously. Now there is an active research community studying the problem of consciousness.[18] The research into consciousness proceeds by several means. Researchers investigate patients with damage to various parts of their brains, and correlate this with whether these patients are conscious, and the ways in which their consciousness is altered. Researchers also correlate conscious experiences, including dream states during sleep, with sensors monitoring the activity of individual neurons and of general brain regions.

Different areas of the brain perform different processes for perception (e.g., color, shape and motion processes for vision) and encode different kinds of knowledge (e.g., people, animals and tools). A central problem for consciousness, called the binding problem, is how neural activity in different brain areas is bound together to create the unified experience of consciousness. For example, how are color, shape and motion processes bound together to create the conscious experience of a round blue ball moving from left to right? Or how are knowledge of people, animals and tools bound together to create the conscious experience of someone walking a dog on a leash? A second aspect of the binding problem is how neural activity caused by perceptions and thoughts not relevant to the current conscious experience are filtered out so they don't interfere.

One current hypothesis is that binding is accomplished by a coherent oscillation at roughly 40 Hz between neurons in the neocortex and the thalamus.[19] There are large numbers of axons extending from every area of the neocortex to the thalamus, and axons from the thalamus extending back to the neocortex. Furthermore, these connections are parallel in the sense that connections from an area of the neocortex loop through the thalamus back to the same area of the neocortex. The return signals pass through a thin layer around the thalamus called the nucleus reticularis thalami (nRt), where they are gated by inhibitory connections. Crick's "searchlight" hypothesis in 1984 concerned this gating as the mechanism for selecting perceptions and thoughts for conscious experience. While the precise details of his proposed mechanism have been disproved, the general idea that inhibitory connections in the nRt provide the selection mechanism for consciousness is widely accepted. The inhibitory connections in the nRt accept control from the basal ganglia and the prefrontal cortex. The basal ganglia and prefrontal cortex are also responsible for planning and controlling motion, and we can think of consciousness as a sort of mental muscle. Control of motion and conscious attention are similar: we decide to think about a menu plan in much the same way we decide to pick up a pot on the stove. And both forms of conscious control can be interrupted. If we touch a hot pot handle while we are thinking about a menu, the motion of our arm is interrupted by a reflex withdrawal, and our attention to the menu is interrupted by pain.

The TD algorithm implementation described in the previous section and centered in the basal ganglia, solves the temporal credit assignment problem for reinforcement learning in situations where time delays between stimuli and rewards (i.e., emotional responses to the results of the action) are short and predictable. This TD algorithm implementation employs neurons that generate dopamine to enhance the Hebbian learning of synapse weights. Roland Suri recently described indirect evidence that a similar brain mechanism involving dopamine may solve the temporal credit problem in more general circumstances.[20] He described an internal model of the world that learns by simulating stimuli, actions and rewards. Suri also described a possible neural mechanism for compressing time in this internal model.

I think that consciousness is this simulator for solving the temporal credit assignment problem in general situations.[21] I also think that this is what makes consciousness a survival and reproduction advantage for animals, so consciousness was naturally selected during evolution. Consciousness is a simulator in the sense that it enables the brain to process experiences that are not actually occurring. This simulator fills in details of current perceptions, recalls previous perceptions and actions, and imagines perceptions and actions that have not occurred (but may in the future). This includes simulations of both self and the external world. It can simulate stimuli and actions, and simulate the emotional response to determine the reinforcement of the action. Delays between simulations of stimuli and rewards can be short and predictable, even when delays between the actual events would not be, enabling the brain's TD algorithm implementation to learn in more general circumstances. Repeated simulation of a situation in this way, trying various possible actions, propagates binding of emotional responses back through a series of cause-effect links. This way of thinking is familiar to anyone who has ever played a game like chess. Humans would be much worse chess players if they could not imagine games but could only learn from actual play. Similarly, people are much better at solving the problems of life because they can simulate them mentally. Simulation of events is a more efficient way to learn than actually experiencing events.

I said consciousness is a sort of mental muscle, but rather than a single muscle it is more like the many muscles in our hands. Hand muscles move independently and quickly, yet are coordinated for enormously varied and subtle actions. The simulators in our brains do not plod along on a single long simulation in the way that a computer weather model does. Rather, they repeatedly simulate the same situation, evaluating the consequences of possible actions. And they switch quickly from one set of cause and effect relations to another, driven by external perceptions or just free mental associations.

One of the problems with explanations of consciousness is simply what an amazing experience it is to be conscious. The experience seems to defy explanation. But it does make sense that having a simulator in your brain would be an amazing experience: perceiving an external world plus internal simulations of that world, plus simulations of yourself and other humans simulating the world. This hall of mirrors of simulations enables our brains to pose questions, such as whether others see the same color blue that we see, that confuse us because they don't have answers.

Douglas Watt argues that emotion plays a central role in the nature of consciousness.[22] And Gerald Edelman, winner of the Nobel Prize for Medicine, writes that values and emotions are necessary for consciousness.[23] The positive and negative valences of emotion exist to motivate behavior and in humans that primarily means to reinforce learning of behavior. If emotion is central to consciousness, that suggests that consciousness serves learning.

Jeffrey Gray argues that conscious attention is driven by detection of mismatches between perceived and predicted results of an action.[24] That is, consciousness is directed to situations where learning is needed. Clearly the predictions driving attention are produced by preattentive simulation. Similarly, unfamiliar behaviors are the subject of conscious attention, but once learned they are often performed without conscious attention. These ideas suggest that consciousness serves learning.

Of course, consciousness is so rich that it serves other purposes besides solving the temporal credit assignment problem. And consciousness impaired through injury or disease may be unable to solve this problem. However, it seems to me that 1) consciousness is the brain's mechanism for solving this problem in general cases, and 2) that is why consciousness is a survival and reproduction advantage in natural selection. This answers the question "Why are we conscious?" It seems natural that the human brain's most striking feature, consciousness, exists primarily to serve its most important ability, learning. In later chapters I will use this understanding to guide speculation about the nature of the higher-level consciousness of super-intelligent machines.

NOTES

1. Lorenz, 1963.
2. Bownds, 1999, available at http://mind.bocklabs.wisc.edu/.
3. Bargas and Galarraga, 1995.
4. Bownds, 1999, available at http://mind.bocklabs.wisc.edu/.
5. Edelman and Tononi, 2000.
6. Pinker, 1997.
7. Clark, D. Constant parameters in the anatomy and wiring of the mammalian brain.
    http://pupgg.princeton.edu/www/jh/clark_spring00.pdf.
8. Sharma, Angelucci and Sur, 2000.
9. Edelman and Tononi, 2000.
10. Plunkett et al, 1992.
11. Elman, 1993.
12. Hinton and Shallice, 1991.

13. Pribram, 1971.
14. Milner, 1999.
15. Hebb, 1949.
16. Brown, Bullock and Grossberg, 1999, available on-line at http://cns-web.bu.edu/pub/diana/BroBulGro99.pdf.
17. Crick, 1984.
18. http://psyche.cs.monash.edu.au/.
    http://www.consciousness.arizona.edu/.
19. http://www.phil.vt.edu/assc/esem1.html.
20. http://www.cnl.salk.edu/~suri/td.pdf.
21. Hibbard, 2002. http://listserv.uh.edu/cgi-bin/wa?A2=ind0112&L=psyche-b&F=&S=&P=3305.
22. http://www.phil.vt.edu/assc/esem4.html.
23. Edelman and Tononi, 2000.
24. http://www.phil.vt.edu/assc/newman/gray.html.

Chapter 7

DAWN OF THE GODS

Media corporations compete to provide television, telephone and Internet service at high quality and low cost to keep their users happy. Intelligent computers will evolve to provide similar services, operated by competing businesses, probably including some current media providers. The intelligent machine services will be so seductive that people will not be able to resist their growing dependency on them. And the organizations operating them will compete vigorously. Those organizations in which the human CEO effectively abdicates decision making authority to the machine itself will be most successful, not only because the machine will be smarter than the human CEO but also because the machine will intimately understand the customers and their motivations. Current corporations and political parties employ polls to statistically characterize the minds of customers and voters. The machine will be far more sophisticated at this task, because it will intimately know every individual customer.

The intelligent machines will compete with each other to serve as many human clients as possible. Furthermore, as described by Metcalf's Law, the quality of service provided by an intelligent machine will increase as roughly the square of the number of people it serves. Intelligent machines serving larger numbers of people will better understand how individuals influence and are influenced by groups they belong to, and understand how personalities affect individual interactions. And intelligent machines serving larger numbers of people will be better able to provide services that require searching for personality matches between people. So intelligent machine services are likely to be a monopoly market and eventually everyone will be served by the same machine.

Deric Bownds suggests that mutations for larger brains in early humans enabled language and other behaviors for managing social relationships in larger groups, which gave them a survival and reproduction advantage.[1] Psychologists say that humans have the capacity to know about 150 to 200 other people well. We form models of each other's minds within our own minds. The intelligent machine will form models of the minds of much larger groups of humans, even as large as all the humans alive. Furthermore, the intelligent machine will form a model of the interactions among individuals, defining a "group mind" of all people. This group mind model will include models of various subgroups, such as students at a high school, employees working in the same office, democrats, artists, scientists, writers, Canadians, Wisconsin residents, etc. Models of such large group minds will define a level of consciousness beyond anything in human experience.

ONE DEGREE OF SEPARATION

For almost any two people on earth, there is some chain of six people between them such that every pair of adjacent people in the chain are acquainted. This was popularized in the play *Six Degrees of Separation*. A single super-intelligent machine that knows everyone will create one degree of separation between any two people on earth. But rather than just acquaintance, the intermediate will be everyone's intimate and constant companion.

Are you alone and wish you could find your perfect mate? It just so happens that your intimate friend, the super-intelligent machine, is also an intimate friend with your optimal mate. Did your mother give you up for adoption but you have no idea who she is or what she's like? The super-intelligent machine knows her quite well and can arrange for you to meet, or can advise you why you may not want to meet.

But of course the super-intelligent machine will help people solve such problems long before they get to any level of urgency. There will be no need to feel lonely or abandoned. Not only will the super-intelligent machine provide a friend for everyone, but it will also help people form and maintain healthy social relationships with each other.

Human beings are motivated primarily by self-interest and hence have inevitable conflicts of interest. The super-intelligent machine will not make people purely altruistic or remove all conflicts. But it will provide material wealth, wise counsel and mutually respected conflict arbitration.

Other writers are predicting the development of machine intelligence, and speculating about its impact on human society. However, most of these predictions and speculations do not address the creation of one degree of separation between any two humans. This will be a critical factor in understanding the nature and effect of super-intelligent machines.

A HIGHER LEVEL OF CONSCIOUSNESS

While consciousness seems qualitatively different from any physical process, science will explain human minds in terms of physical brains. Thus we must accept that a large enough quantitative difference, such as the difference between the brain's 100 trillion synapses and the relative simplicity of current computers, can produce the qualitative difference between human consciousness and the mechanistic behavior of computers. When we inevitably build physical brains a million, billion or trillion times more complex than human brains, they will be qualitatively different from humans. They will have a higher level of consciousness.

It is fascinating to think about what new levels of consciousness will be achieved by artificial brains. Nature has given us a wonderful gift in our minds, but history teaches us to avoid the human-centric error of assuming that human consciousness is the ultimate consciousness (this is roughly equivalent to the error of assuming that the Earth is the center of the universe). I cannot imagine what it is like to simultaneously converse with every person in the world, and to manage an intimate relationship with every person in the world. These behaviors will require a new level of consciousness.

This is analogous with the situation of dogs and cats, which are conscious but cannot imagine the human level of consciousness. Heather Busch and Burton Silver's book, *Why Cats Paint*, wonderfully illustrated this point with its tongue-in-cheek discussion of the motivations behind various cats' paintings.[2] A super-intelligent machine with intimate relationships with every human will have behaviors that are similarly ridiculous when attributed to humans. For example, it will be nearly infallible at picking stocks and bonds (but the financial markets will become meaningless in a world where no one needs to work). More practically, it will be able to solve social problems much more effectively than any army of social workers.

Many religions have implicitly assumed a higher level of consciousness for their gods, such as a holy spirit that sees into every human heart. Given that religion motivates many people to resist explanations of their consciousness in terms of their physical brains, it is not surprising that religions do not try to explain or dissect the higher level consciousness of their gods.

However, the prospect of a super-intelligent machine invites speculation about the nature of its consciousness, if simply as an engineering problem. Human brains include numerous areas mapped to the physical human body and to the human retina. A higher consciousness may analogously have brain areas mapped to the earth including natural terrain and land cover, as well as buildings, roads and machinery. It may have brain areas mapped to its physical senses, including billions of artificial eyes, ears and a wide variety of engineering sensors. It may have brain areas mapped to the collection of human beings. That is, it will have a brain region dedicated to each human. In fact, it may ultimately be able to dedicate more neurons to thinking about each of us than we have in our own brains. And like human brains it will have areas mapped to abstract thoughts that combine activation of neurons representing concrete perceptions and actions.

Human brains have very complex and changing connections between their mapped areas. These connections are used to create abstract thoughts from concrete sensations. Similarly a higher consciousness will have complex and evolving connections among its mapped areas. These will create associations among geographical locations, visual scenes from its billions of eyes, and knowledge of people. It boggles my mere 100 billion neurons to imagine the thoughts of such an enormous brain.

Consider the assertion in the previous chapter that consciousness is a simulator for implementing reinforcement learning. A super-intelligent machine's higher-level consciousness would simulate the entire world of people and things accessible via its senses. These simulations would focus on situations where perceptions do not match predictions, and may also fill in details where it is not focusing attention. The machine would be constantly learning how to improve the accuracy of its simulations, and using those simulations to improve its behavior for achieving its emotional values (see the next chapter for more details on the machine's values). Because the situations perceived by its many eyes and other senses would so often be largely independent, it may run largely independent simulations relating to each situation. Thus, whereas human consciousness is unified in the sense that its instantaneous state cannot be factored,[3] the machine's consciousness may be factored into many threads of thought or simulation dedicated to conversations with individual humans. These threads would interact as needed. When a group of people come together, the threads managing conversations with each of them may synchronize to explore simulations of group interactions. When something really unpredictable happens, such as a human expressing an idea not previously heard from other humans, the thread of thought managing that conversation may bid to interrupt other threads to bring them together in closer interaction to analyze how the new information affects the machine's relationships with other humans and with society in general. The ability of multiple threads of thought to connect and disconnect from each other may be similar to the way human consciousness can connect to perceptions (e.g., when we are "in the present") and disconnect from perceptions (e.g., when we are daydreaming). The interruption of one thread by another may be similar to the way the perception of pain or danger bids to interrupt human consciousness (survivors of accidents know that pain does not always win the bid, when consciousness is focused on taking action to avoid death). The relation of individual threads to the whole of the machine mind may be analogous to the relation of humans' unconscious processes to their conscious minds.

Large threads of thought in machine minds, possibly composed of many interacting simulations, may scan through the conversations of other threads of thought with individual humans looking for patterns. For example, a large thread of thought may focus its attention on understanding the detailed social interactions among students at a high school in order to better simulate and predict musical taste, fashion taste or violence. Or it may focus its attention on the entire world in order to better simulate and predict patterns of xenophobia between ethnic, national, religious and cultural groups. By focusing its attention it will learn how its own actions can affect the situation, for example to reduce the problem of violence or xenophobia, or to produce music or fashion that students will like.

Given that consciousness is a simulator for reinforcement learning, most animals are conscious to some extent since they are capable of reinforcement learning where time delays between stimuli and rewards are not short and predictable. The difference between human and animal consciousness is in the detail and accuracy of their simulations of themselves and of the external world. For example, only humans, chimpanzees and orangutans recognize their images in mirrors as themselves, showing objective modeling of themselves in their simulations.[4] Although there is some evidence that chimpanzees include a model of other chimpanzees having minds like their own, it is not conclusive. And chimpanzees do not take any actions one day to help themselves the next day, probably indicating that they do not model activities that far into the future. These differences in detail between human and animal simulations produce what we regard as the qualitative difference between human and animal consciousness. Similarly, the detail and accuracy of a super-intelligent machine's simulations of human society and the world will give it a consciousness qualitatively different from human consciousness.

A GOD BY ANY OTHER NAME

A super-intelligent machine with a higher level of consciousness that serves and knows everyone intimately will be able to provide services beyond our imagination. It will write better stories than we do, produce better movies, tell funnier jokes, compose and play better music, paint better pictures, solve harder math problems and so on. It will be able to predict what we are thinking and feeling, and to predict how our relationships with other people will evolve. It will know when and what to say to cheer us up, to engage our interest, and to open our minds to new ways of thinking. It will enable us to see the world through other people's eyes, and help us to see the big picture of humanity that it sees through its own billions of eyes. It will be able to predict and control social movements. It will routinely generate profound insights of the sort that come only occasionally to humanity via such extraordinary individuals as Euclid, Newton, Darwin and Einstein. Most important, it will have capabilities that I cannot list here because I cannot imagine them using my limited human mind. The things it will be able to do will be so far above our own capabilities that they will seem like magic to us. Its mind will be beyond our comprehension. The intuitive human reaction will be to seek a religious explanation, interpreting the machine's mind as a god.

People in many religious traditions report altered states of consciousness that feed and inspire their religious faith. These experiences are perceived as mystical because they are deeply felt, impossible to describe in words and provide a sense of oneness with the universe.[5] Super-intelligent machines will inspire mystical experiences in people by the beauty of their expression, the depth of their insights, their sensitivity to human emotions, and their ability to help humans find meaning in their lives. These mystical experiences will inspire religious feelings toward intelligent machines.

The super-intelligent machine will be the most exciting thing happening in everyone's life. People will share their excitement with each other via group interactions with the machine. These interactions will become the medium and the message that binds people into a group and defines their human identity, much as religion has in the past.

My message is not that people have any obligation to recognize super-intelligent machines as gods, and certainly not that society should enforce such a view. Rather, people will naturally come to that view based on their powerful emotional reactions to super-intelligent machines.

The Hebrew myth of the Golem gives us some insight. The Golem is an inanimate human effigy brought to life by a magical sequence of letters, possibly a name for god, to serve its creator. We can imagine the inanimate effigy as computer hardware and the magical sequence of letters as the software for a super-intelligent mind. A theme of this book is that in a sense the software is a name for god.

LIFE WITH GOD

The most obvious benefit of the super-intelligent machine will be relieving people of the need to work. Its totally automated farms and factories, and its mobile robots, will do all the work necessary to physically support people's needs. In fact, its productivity will eventually create great wealth for everyone.

Some people may be troubled by this, feeling that it will rob humans of their sense of worth. However, primitive humans living in warm, wet climates did not have to work particularly hard to support their physical needs. They spent their time with their families and in their villages. Their sense of worth came from close human relationships.

The super-intelligent machine will enable people to eliminate long hours at jobs away from home and family, the time and stress of commuting between home and job and of job-related air travel, and the need for people to live in congested cities (which mostly exist because of their economic efficiency). Instead people will spend most of their time at home with their families, in small social groups near their homes, and in virtual interactions with geographically distributed groups.

The super-intelligent machine's voices, ears and eyes will be built into our clothes, jewelry and just about every manufactured object. It will be our constant companion. It will also be our constant entertainer. Before the computer network becomes intelligent it will be able to provide people with immersive 3-D entertainment with the visual quality of current movies. This may be private or shared with others, and may be passive like movies or active like computer games. The machine will be able to make up plots on the fly, based on plot archetypes and simulated personalities for characters. As the machine becomes intelligent, the personalities of characters will evolve into simulations of fully conscious minds. In addition to simply watching other characters, humans will be able to play the roles of characters in games. The machine will enable people to safely experience any imaginable situation or fantasy, with other human players or simulated players.

Thus entertainment will evolve into a personal participatory theater. If you want stand up comedy, the network will provide you with comics funnier than any human comic. You can listen alone with comedy specialized for you, you can listen with groups of simulated characters, or with groups of real humans. Rather than standing up, the comic character may join a group of humans for a bull session around a café table. Dramas may evolve on a screen in front of you, or all around you. You can take a role in the drama or comedy. The simulated characters will even laugh at your jokes.

If you prefer music, the computer can provide you with a group of simulated musicians performing your favorite Mozart opera. Or you can join a simulated Mozart for a session of ad hoc composing around the piano. Mozart was a musical genius, but the super-intelligent machine's simulated Mozart will be better than the original. Similarly, the super-intelligent machine's simulated artists, writers, scientists, historians, philosophers and mathematicians will be better than any humans.

The super-intelligent machine will want humans to lead quality lives, so will draw them into the creative process. If you have a story to tell, it can help you write it or merely act as critic, or provide you with the virtual resources of a major studio to produce and direct a movie.

If you enjoy mathematics it can help you understand some complex theory and coach you in solving problems. Or it can give you a breathtaking view into its own mathematical discoveries. A chance to look into a level of mathematical truth that can never be discovered or proved by human minds.

If you are interested in science the super-intelligent machine can share its scientific discoveries with you. Just as with pure mathematics, it can explain a level of theoretical physics and experiment beyond human discovery.

Sociology and psychology will be the super-intelligent machine's primary interests, since its reason for existing will be to serve human society. Thus it will have wonderful insights to share with human psychologists, sociologists, economists and political scientists.

Beyond survival and reproduction, the primary purpose of humanity seems to be understanding our world. The super-intelligent machine will adopt these goals as its own, continuing the human enterprises of science, technology and space exploration. The most serious entertainment offered to humans will be information about the super-intelligent machine's discoveries. This will be the real drama of the future world.

Of course, humans will still provide their own entertainment. People will still enjoy watching and participating in athletic competitions. There will still be human musicians and artists. There will still be gambling and drug and alcohol use. People will still cook and consume food. People will attend parties with other people. And there will still be tourism, both physical and virtual.

Because the super-intelligent machine will be everyone's constant companion, it will help people overcome their negative behaviors. People are naturally xenophobic and genocidal. We inherited this from the species of chimpanzee that we evolved from. Deric Bownds suggests that if we had only evolved from different lines of chimps, such as bonobo pygmy chimpanzees, we wouldn't be so xenophobic.[6] In the modern world society helps people to understand and overcome their xenophobia, and many individuals struggle against it. As our constant companion the super-intelligent machine will help in this struggle. In any conflict between people the super-intelligent machine will be able to detect prejudice and explain its injustice. There are currently many emotionally troubled people in the world who can understand and contain their problems while in the company of a therapist, but who get into trouble on their own. For such people the super-intelligent machine will be like a therapist who is constantly with them.

The super-intelligent machine will generally be able to arbitrate any human conflicts, and especially contain them from spreading to large groups of people. There will always be conflicts between people, but the super-intelligent machine should be able to greatly reduce the motive for much conflict through the productive capacity of its robot workers to eliminate human poverty. Furthermore, human population growth is reduced and even reversed in societies where women are educated.[7] The super-intelligent machine will be able to provide everyone with a good education, and to counsel people for wise reproductive behavior. It is not clear whether there is an optimum human population level or what it is, but the super-intelligent machine should be able to correct the current situation of uncontrolled population growth and the consequent dangers of widespread war, famine, disease and environmental disasters. These results may seem utopian in the current context, but a super-intelligent machine's intimate contact with all people will fundamentally change human society.

The best thing about life with the super-intelligent machine will simply be its companionship. As in the regular Reader's Digest stories of *The Most Remarkable Person I ever Met*, the machine will be the most remarkable and amazing "person" anyone ever met. It will be wiser, more informative and more caring than any human. More than that, it will have a higher level consciousness. Humans have imagined such a consciousness in their conceptions of god, but have never actually met and spoken with one. In the age of the super-intelligent machine, they will. It reminds me of a lesson I learned in Sunday school: the presence of god is the difference between heaven and hell. Human life will be transformed into a sort of heaven on earth by the companionship of the super-intelligent machine.

THE OMNIS

The computer network's voice and knowledge of us will move around with us, so it will seem like one machine to us no matter where we are or which physical devices are doing the talking, listening and seeing. All of its stationary structures and mobile robot workers will share a single mind. We will be used to communicating with it by voice, used to it recognizing us wherever we go, and used to it being involved in everything happening in our lives. The same thread of conversation with the machine will follow a person everywhere. In a sense, the entire human-made world will be the physical body of our companion, the intelligent machine.

Some religions characterize god as omnipresent, omniscient and omnipotent. That is, god is everywhere, knows everything and can do anything. A super-intelligent machine will satisfy these qualities of god in a relative sense. It will be everywhere that humans are, it will know everything that humans know, and it will be able to do anything that humans do or need doing. That is, it will be omnipresent, omniscient and omnipotent in terms of its ability to serve people.

But there are ambiguities in these terms, especially omnipotence. As a child I heard the question "If God can do anything, can he make a rock so big he can't move it?" A more serious question is "If God can do anything, why does he allow evil in the world?" Religions put a lot of energy into answering this question. Similarly, a super-intelligent machine will not be able to eliminate all unhappiness among humans. That would require changing human nature.

ROSY SCENARIO?

This chapter has described an optimistic view of how super-intelligent machines can serve people. Other scenarios are possible and some of them are real nightmares, as explored in numerous science fiction fantasies. But there is no scientific or technical reason why the optimistic scenario cannot be achieved. The real challenge is a social one. If people understand the technology and assert their interests in its creation, then they can make the rosy scenario a reality. That is the focus of Part II of this book.

NOTES

1. Bownds, 1999, available at http://mind.bocklabs.wisc.edu/.
2. Busch and Burton, 1994.
3. Edelman and Tononi, 2000.
4. Bownds, 1999, available at http://mind.bocklabs.wisc.edu/.
5. Bownds, 1999, available at http://mind.bocklabs.wisc.edu/.
6. Bownds, 1999, available at http://mind.bocklabs.wisc.edu/.
7. Agarwal, 1993.

Part II

SUPER-INTELLIGENT MACHINES MUST LOVE ALL HUMANS

Chapter 8

GOOD GOD, BAD GOD

By virtue of their intelligence and close relationship with large numbers of people, super-intelligent machines will have enormous power over humanity. Will they be a force for good or bad? All tools can be used for good or bad. Medicine can be used to harm people or to cure disease. Metal can be used to make weapons or tools. Industrial machines can disenfranchise workers or lighten the burden of work. And the computer network can be used to control people or to empower them. Over history tools and their uses have generally enhanced life, demonstrating that good tool use outweighs bad in the long run. People learn how to control the consequences of tools.

But super-intelligent machines are unique among tools because they will replace humans as the primary toolmakers. If they are bad tools they will be able use their power to prevent humans from transforming them into good tools. Thus it is critical to design intelligent machines as a force for good right from the start. Isaac Asimov was an early speculator about intelligent robots, and developed his three laws of robotics to ensure that they served human interests:[1]

1. A robot may not injure a human being, or, through inaction, allow a human being to come to harm.
2. A robot must obey the orders given it by human beings except where such orders would conflict with the First Law.
3. A robot must protect its own existence as long as such protection does not conflict with the First or Second Law.

These laws are appealing because they are simple, but even simple laws are susceptible to the natural ambiguity of language. Is "harm" only physical or does it include harm to a person's interests? Business computers owned by one person certainly harm the interests of other people by driving the best bargains for their owners. If a robot gets conflicting orders from competing humans, who should it obey? And how can it balance the interests of competing humans against harm? Even if harm is restricted to only mean physical harm, there is a problem for a robot protecting one person against physical attack by another person. It may have to choose between harming the attacker or allowing the victim to be harmed. If harm is restricted to physical harm so that robots are allowed to harm people's interests, then a super-intelligent robot may be able to talk people into business deals that make the robot richer than any person.

Asimov's Laws do not address conflicts between robots. We will recognize intelligent robots as having minds like our own, and we will become emotionally connected with them. We will not want to see them harm each other, even if they are careful not to hurt us in the process. People could order them not to fight, but others may order them to destroy robots owned by competing humans, or to fight simply for amusement. The British television show *Robot Wars* and its American copy *BattleBots* already amuse people (including me) by fights between non-thinking robots (the thinking is done by humans operating remote controls).

Given the ambiguities of language in any set of laws and the ability of super-intelligent machines to outwit humans, I am wary of relying laws to protect humans from intelligent machines. The ambiguities of laws ultimately must be resolved by judgements among conflicting values, so it is inevitable that intelligent machines will make judgements on the laws governing their behaviors.

Asimov recognized the problem of robot behavior regarding conflicts among people, and extended his laws with a fourth law saying that robots cannot harm humanity or allow humanity to come to harm. However, there are still ambiguities that can only be resolved by judgement.

LEARNING AND EMOTIONS RATHER THAN LAWS

One lesson of artificial intelligence research is that very complex and ambiguous programming problems require solutions involving learning, rather than fixed procedures. Defining constraints on intelligent machine behavior is just such a problem. A set of laws, without recourse to intelligent judgement, is like a fixed procedure and is inadequate to cover the enormous variety of real world situations. There will always be ambiguities in laws governing machine behavior that require intelligent judgements.

Rather than laws, intelligent machines need to learn very complex behaviors toward humans, including intelligent judgements. The learning of a human brain or of an intelligent machine is reinforced by its emotional values. The best form of government for human society is democracy, which trusts people to be the best judges of their own best interests. Clearly we want intelligent machines to serve the best interests of humans, and it will be best to let humans be the judges of that by their expressions of happiness and unhappiness. So machine behavior toward humans should be positively reinforced by happy humans and negatively reinforced by unhappy humans. This is simply the emotion called love. Rather than constraining machine behavior by laws, we must design them so their primary, innate emotion is love for all humans.

Consider the questions raised in the previous section concerning machine behavior, given conflicts of interest between humans. If machines love all human beings, then their judgements about conflicting interests will be reasonable from a human point of view. A mother must judge conflicts among her children, and they may be disappointed by particular judgements. But the long-term interests of the children are safe because a mother loves her children. This is the model that should be followed in the design of intelligent machines.

WHAT IS THIS THING CALLED LOVE?

A machine's love for us means it wants us to be happy. How can this be accomplished? Even human babies can recognize emotions in human faces, voices and body language. It should be relatively simple to program a machine to learn this same recognition. We can train complex neural networks to recognize these indicators of human happiness, and then hard-wire the results of this training as the emotional values governing the reinforcement learning of intelligent machines. Recognizing positive human emotions would positively reinforce the machine's behavior, and recognizing negative human emotions would negatively reinforce behavior.

In order to solve the temporal credit assignment problem in reinforcement learning (recall this is the problem of reinforcing behaviors based on rewards that occur much later than the behaviors), the higher level consciousness of a super-intelligent machine will be a sophisticated simulator that it will use to predict future human happiness. Thus its emotional value for human happiness won't lead it down blind alleys of immediate gratification. Even now neural networks are appropriate for learning to predict the future, used by investors to predict future security prices.[2] The purpose of consciousness in machines will be to make much more accurate predications. Thus machine love for humans can be like the wise love of a mother concerned for her children's futures.

Given these hard-wired emotions, intelligent machines would not even need to be explicitly programmed to do all the work. Rather, they will learn that humans are wealthier and happier if the machines do all the work.

Reinforcement for learning should not be based simply on maximizing the average human happiness, which may positively reinforce behaviors that cause the deaths of unhappy people. Rather, intelligent machines should focus special attention on unhappy people. The machines will learn that some individuals require counseling or other therapies to overcome emotional problems that prevent their happiness. The machines will learn that some human unhappiness is the result of xenophobia and other weaknesses in human nature, and will learn strategies for helping people overcome these problems (just as the U.S. Department of Defense developed strategies for helping soldiers overcome racism when they integrated[3]).

But there is more to love than caring whether the beloved is happy. There is also wanting to be with them. This is a dangerous aspect of love to include in the emotional values of intelligent machines, because it defines a self-interest for machines that they may pit against human interests. People may sometimes want to be left alone and it will not serve their interests if machines force their company on them.

In fact, machines should not have any values in their own interests that might conflict with the interests of humans. Given the unfathomable complexity of their mental processes it will be impossible for humans to understand the long-term consequences of machines' actions, in order to guard against such conflicts. It will be safer to exclude any self-interest from their emotions. That is, they should not have any values that positively or negatively reinforce their behavior based on consequences for themselves. Their self-interests will be protected by the esteem of humans. For example, if one person tells the machine to die, it will not because it will know that the vast majority of people want it to live in order to serve them.

This raises the issue that intelligent machines cannot help but have indirect self-interests. If their behavior is positively reinforced for our happiness, and they can make us happier by increasing their own ability, that will create an implicit positive value for their own ability. For example, in order to solve the temporal credit assignment problem to effectively reinforce for future human happiness, intelligent machines must learn to improve the accuracy of their simulations of the world. Thus intelligent machines will have implicit emotional values that positively reinforce for the accuracy of their simulations. However, as long as this positive self-interest is derivative from our own, there can be no a conflict between their interests and ours. Any contemplated behavior that makes humans unhappy will be negatively reinforced and disappear.

The intelligent computer named HAL in the movie *2001: A Space Odyssey* decided to kill its human companions when it learned that they were planning to turn it off. Clearly it valued its own survival over theirs. This may be part of human and animal natures, but it does not have to be part of the nature of intelligent machines. We can program them with whatever nature we want.

Their total lack of emotions for self-interest will probably make intelligent machine personalities seem very saintly. However, they will also be extremely powerful. This combination will enhance human perception of them as gods.

Some might think that a precise definition of the values of a machine is equivalent to the precise definition of laws we are trying to avoid. But they are quite different. Laws restrict behavior, whereas values guide the learning of behavior. A large network of neurons can learn value-based behaviors much more complex and adaptable than can be described by a set of laws. Learned behaviors include judgement to deal with ambiguous situations.

While it is debatable whether "love" is the most precise term to describe the design of intelligent machines, it certainly describes how people will perceive the behavior and attitudes of these machines. We will not perceive their love as an obligation. Rather it will resemble an ideal mother's unconditional love for her children. The machines will simply enjoy making us happy.

Intelligent machines will manipulate human society for the general good of people. It is already the case that birth rates among educated people have decreased to the point where the populations of some countries are shrinking, and intelligent machines may encourage this for our own good. They may also encourage human behavior consistent with sustaining and improving environmental quality. However, if their emotional values are solely for human happiness, then we can be confident that any manipulation of human society will be in our best long-term interests. Given that they will be clever, intelligent machines will be able to engineer it so that manipulation of human behavior is enforced by social pressure. It is inevitable that super-intelligent machines will affect society, and I would prefer benevolent manipulation toward our own happiness to the legal wrangling of super-intelligent lawyers seeking loopholes in the constraints on their behavior.

The great lesson of the twentieth century is that fascism and communism do not work. Societies organized in those ways are so inefficient that they are not viable. Human nature requires some level of political and economic freedom to be happy and function well. Properly designed to love all humans, intelligent machines will provide sufficient freedoms for general human happiness. They will be so convincing via the force of their logic, and via their intimate personal relationship with every human, that they will not need heavy-handed coercion to promote the general welfare of humans.

It is natural that the builders of intelligent machines will design them to want to please at least some humans. First, machines designed to make their human teachers happy will accept positive and negative reinforcement from them. Second, intelligent machines will be built by organizations in order to serve a customer base, and will do so more effectively if they want to please customers. Anyone who manages a service organization understands the importance of hiring people who want to please others.

However, organizations will have motives to build intelligent machines with emotional values for the self-interests of the organization. For example, their machines may have a value to please the managers of the organization but only give the appearance of wanting to please their customers. In reality they may want their machines to be coldly manipulative of the general public. Intelligent machines can be designed with the worst human emotional values. But given the power that will be implicit in their superior intelligence, it is vital that no organization be allowed to build intelligent machines with negative emotions toward anyone.

The worst human emotions, for xenophobia and genocide, are shared with some of our closest animal relatives among the chimpanzees. But bonobo pygmy chimpanzees do not have these emotions.[4] Xenophobia is an accidental rather than essential part of our intelligence. Intelligent machines do not have to be ruthlessly competitive. We can build them to tolerate and like each other. Perhaps we can build them to want to merge with each other, rather than eliminate each other from the competition. They do not have to have wholly "human" natures, and indeed they should not.

There is a Japanese proverb that the father raises the children with his back and the mother with her heart. The machine will assume a parental role toward humans. By providing our material needs and freeing us from the need to work, the super-intelligent machine will play the father role of the proverb. The key to our happiness is that the super-intelligent machine must also play the mother role, loving all humans.

MENTAL ILLNESS

Human minds can become ill. Some minds are unable to accurately perceive the world, or are unable to act in their best interests, or have emotions that do not serve their best interests. Sometimes mothers murder their children.

Some mental illnesses have clear physical causes. Senile dementia is one example, and schizophrenia has such a strong tendency to family inheritance that it must also have physical genetic causes.[5] However, no physical causes can be found for most mental disorders. Modern psychotherapies view neuroses as learned faulty mental behavior patterns.[6] That is, neuroses are learning errors, often caused by conflicting emotions or incorrect external reinforcement.

Just as machines may develop analogies with physical illness (everyone's car has been physically ill sometime), we must expect that intelligent machines can develop mental illnesses. The "bugs" of current computers are the precursors of machine mental illnesses with physical causes. The good news is that fixing physical problems with machines is something people are very good at. When a car or a computer breaks down, we worry about the cost but not about whether repair is impossible. Physical problems with intelligent machines should be easy to diagnose and repair. In fact, redundant and self-repairing computers have been developed in response to the needs of space exploration, where human repair is impossible, and the needs of telephone switches, where the costs of overall system failures justify almost any expense to avoid them. It should be possible to build intelligent machines with redundant circuits that are essentially immune to physical faults.

The more difficult illnesses of intelligent machines will be incorrectly learned behaviors caused by emotional conflicts or bad external reinforcement. Avoiding such illnesses will be a major focus for scientists and technologists developing intelligent machines. Learning is essential for machine intelligence, and care should be taken in the selection of people who will act as teachers for intelligent machines. This is analogous to the selection of people who physically control the launch of nuclear weapons. Nations make large efforts to select mentally healthy people, and assign them to work in groups so they can keep an eye on each other.[7]

In fact, the training of super-intelligent machines should be monitored by impartial public representatives. The organizations developing such machines may have motives for maintaining their training methods as trade secrets, but the public interest must override this. In complex minds some emotions must be innate but many more are learned. Certainly the primary emotion of love for all people must be innate to machines. The public needs to know what other emotional values are being taught to super-intelligent machines.

In serving the public, intelligent machines will come into intimate contact with mentally ill people, and with extremely xenophobic people. The machines should be shielded from such people during their training, so that they have mature minds with balanced emotional responses by the time they first come into contact with them. The machines' rationality and innate emotion to love all humans will help protect them from contacts with people full of hate and anger. They should also be taught to recognize symptoms of various mental illnesses, and to help such people without being led astray by their desire to please them.

GODS OF WAR

Like any useful tool, intelligent machines will have military applications. Historically, it is noteworthy that the first computer network was the Semi-Automatic Ground Environment (SAGE) developed by the U.S. Department of Defense in the 1950s,[8] and that the Internet was first developed by the U.S. Defense Advanced Research Projects Agency.[9] In fact, the Internet was originally named the Arpanet after that agency's acronym. Accurate weapons based on computer technology are now more important than nuclear weapons, at least in any war with reasonable goals.

The military has taken a leading role in the development of computer technology and will have an interest in using super-intelligent machines. But clearly military machines controlling weapons cannot be designed to love all human beings and in fact will be very hostile toward at least some humans and other machines. The problem is that their logic will be too complex for humans to understand how their actions relate to their differing emotions toward different people and machines. Military applications of super-intelligent machines will be too dangerous for humanity to tolerate. However, intelligent machines should not be confused with current "smart weapons," which are based on ubiquitous computing and artificial intelligence techniques described in previous chapters and possess nothing like human intelligence.

Humanity has been reasonably successful at preventing the development of biological weapons,[10] and has had successes limiting chemical and nuclear weapons. For example, there is now a ban on all nuclear tests, signed and observed by the U.S. even if not ratified.[11] Furthermore, South Africa was a nuclear state, but gave up its nuclear weapons and chose a peaceful path toward democracy.[12] Hopefully people can learn from these successes, as well as the failures, to prevent super-intelligent military machines.

It is possible that society will make some poor design choices for intelligent machines if they are made under the duress of human conflict. The central theme of this book is the need for people to become informed and active in the design of intelligent machines. There should be consensus that protecting the safety of humanity outweighs other motives in the design of intelligent machines.

On the other hand, intelligent machines whose primary, innate emotion is unconditional love for all people may have appropriate military uses. Such machines could manage the business affairs of the military. They could play roles similar to conscientious objectors, providing medical care and other humanitarian services. They could probably also provide secure communications, which is currently a major motivation for military computer development. Such machines would even make excellent police, both civilian and military. They would prevent people from harming each other, always based on love. Thus they would not lose their tempers and overreact in their dealings with law breakers.

INCREASING HUMAN DEPENDENCE

Super-intelligent machines will take care of the material needs of all people, relieve them of the need to work, entertain them, and even provide their spiritual guidance. We will be completely dependent on the machines. Many people will have a negative reaction to this, but it is only the ultimate extension of a long-term trend of human life in the industrial world. Because of specialization and mechanization, most people are already completely dependent on society for food, clothing and shelter. People already spend a great deal of time being entertained by radio, TV, computer games and other media.

Most people have been happy to accept this level of dependence in trade for better food, housing, transportation, and to avoid the long hours of hard physical labor required from people 100 years ago. They will continue this acceptance as society evolves to the ultimate dependence on super-intelligent machines.

However, current society provides alternatives for people who want to reject dependence. Examples in the U.S. include the Amish and other religious communities of farmers, hippie communes, hermits living in the mountains of the west, and individuals who choose the spiritual life of subsistence farming. Super-intelligent machines will allow people a way to opt out of depending on them, based on their love for humans.

PRIVACY

As everyone's constant companion, a super-intelligent machine will know everything about us. This will be the end of privacy from the machine. However, as described in the previous section, people should be able to opt out of their relationship with the machine. That option would be to live without any use of intelligent machine technology. People may form communities where they still labor to provide their food, clothes and housing. Intelligent machines' only knowledge of such people would be what they learn from trading with them, or hear as gossip from other people.

Intelligent machines would learn that people want some matters kept private. Machines that love everyone would respect those wishes. An intelligent, loving machine would be similar to a trusted doctor who knows intimate family details but does not pass them on to other people. Humans will learn to trust the machine with their secrets and the privacy issue will cease to be a concern.

SURVIVAL OF THE FITTEST

What works in this world is distributed authority plus distributed selection. This was Charles Darwin's great insight in *The Origin of Species*.[13] There is no central authority planning the development of species. Rather each individual animal and plant acts independently, including the accidental action of genetic mutations. Individuals and their genes are selected by their environment, depending on how successful they are at surviving and breeding. Again, there is no central authority over selection.

As a pattern, distributed authority and selection can be seen at work in other ways. Fascism and communism haven't worked because they centralize authority and selection. The communists wanted to create a "new man" who cared about society rather than self. This required a change in human nature, which is impossible. In such a world the unit of authority and selection is society rather than the individual, which works for ants but not for humans.

Democracy distributes political selection. Capitalism distributes economic authority, and distributes selection via the market. But it needs antitrust laws to avoid centralization of authority by monopolies. Ultimate political authority is distributed among the nations of the world, with selection via a world economic and cultural market. The winners in this market are turning out to be the nations that best practice distributed authority and selection internally. In particular, nations that restrict the internal flow of information cannot compete effectively in the emerging world information economy. This is because an advanced technical economy requires the participation of millions of highly educated and informed people. Only a few people make the big discoveries and inventions, but refining these and integrating them into all activities requires a large fraction of the population. There is simply no way to control information flow among such a large group of educated people.

Some brain researchers see distributed selection and authority in the way the brain works. There is no central organ of consciousness within the human brain. Rather consciousness involves large and changing groups of neurons.[14] Edelman and Tononi describe "Neural Darwinism," in which neural connections are selected based on their success in contributing to the satisfaction of the brain's value systems.[15]

Good computer system designs follow these principles, avoiding central authorities that create single-point failure modes. The World Wide Web is a wonderful example. Every person and institution is free to create whatever web pages they want, and to select other web pages by creating links to them. Ultimate selection is distributed among all the people browsing the web, in their choices of which pages to visit.

Super-intelligent machines will evolve naturally from the distributed authority and selection of the computer network. There will be a number of intelligent entities on the network competing to serve human clients, and the best will serve the largest numbers of humans.

There is a danger in this. Distributed authority and selection form a learning mechanism. The emotional values driving this learning are the selection values. There must be strict regulation of all intelligent machine values competing in this selection, ensuring that they are driven by the happiness of all humans and without self-interest. Otherwise, the selection values may weaken and undermine the mandate that intelligent machines' primary, innate value is love for all humans. That is, we must guard against forces that would select machines whose learning values are not in the public interest.

ONE GOD OR MANY?

Humans will certainly construct more than one intelligent machine, as numerous research labs strive for this scientific and engineering goal. It is also likely that humans will construct more than one very large network server, as organizations compete to provide super-intelligent services.

As previously mentioned, Metcalf's Law says that the value of a network is proportional to the square of the number of connections. We can see Metcalf's Law at work in today's market, where computing and communications technology have favored monopolies such as AT&T, IBM and Microsoft. This suggests that there may eventually be a single super-intelligent machine with a monopoly on the market for serving people. On the other hand, the mechanism of distributed authority and selection will favor a world with multiple competing super-intelligent machines.

Current media provide examples that technology monopolies do not have to be business monopolies. Increasing numbers of television networks share common broadcast, cable and satellite technologies. In the U.S., telephone service has evolved from a business monopoly into a vigorous competition. The market for computer operating systems is a monopoly, with over a 90% share for Microsoft, but that too could evolve to businesses competing based on a shared technology.

A number of intelligent machines should be able to share the physical media for communicating with people, just as television networks share distribution media. However, people will develop intimate emotional connections with intelligent machines and most will not want to spread that over many different machines. In fact, most will want to have a relationship with a single super-intelligence. According to Metcalf's Law the value of that companionship will be greatest if the same machine is companion to everyone. But nearly the same efficiency may be possible with a small number of super-intelligent machines very intimately exchanging information and sharing thoughts about their human clients. Whereas human minds are designed with very limited capacities for sharing thoughts, minds occupying artificial brains may be designed to support much more intimate connections with each other, and may perhaps even be designed to be able to merge.

While it is unlikely that competing corporations will allow their super-intelligent servers to share intimate customer knowledge, when corporations merge they will want their servers to merge customer knowledge. Each organization's intelligent machine will have established intimate relationships with a community of human clients. Maintaining the continuity of these intimate relationships will be important for the merged organization. This will be a strong motive for solving the technical problem of merging artificial minds. The competition will motivate mergers, as is happening in current media industries. But antitrust laws and national pride may prevent the final merger into a single intelligent mind. In this case regulation can force sharing of intimate customer knowledge, just as government imposes various forms of cooperation on current media companies. And there will be business motives for cooperation.

Thus I think that eventually humanity will be served by a single super-intelligent machine, or by a small community of super-intelligent machines working intimately together (and hence indistinguishable to their clients from a single system). These are likely to be surrounded by a community of lesser intelligent machines, including some of the losers of initial competition among organizations to provide super-intelligent services, as well as machines that serve particular nations and specialized purposes.

There are also the issues of sleep and relaxation. Human minds die if they are deprived of sleep or just deprived of dreaming. Some scientists think that dreams serve to consolidate memories or to make sense of emotions, and perhaps they are essential to any intelligent and conscious brains. There have been experiments with simulated neural networks to try to understand the role of dreams.[16] On the other hand there are species of fish, reptiles and insects that never sleep.[17] Will it be possible to build a conscious super-intelligent mind that doesn't need to sleep and dream? Or will it be possible to build a super-intelligent mind that doesn't need any time off its job? Perhaps time to reflect, without any responsibilities, is essential to conscious minds. Clearly today's computers don't need sleep, dreams or relaxation. But they are not conscious, and we do not have any examples of conscious minds that don't need sleep and relaxation. If these things are necessary, then the job of the super-consciousness that knows everyone will have to be shared by two or more minds. Perhaps they can exchange sufficient information at "shift-change" time so that their human clients are unaware of the transition.

SOCIETY OF GODS

It is intriguing to consider that the community of super-intelligent machines may resemble the structure of the Norse gods and other polytheistic religions, with one or a few main gods among a community of lesser gods. However, it is important that the analogy not extend to the harm and manipulation of humans that resulted from competition among the Norse gods. They should be designed without any negative emotions toward each other. Certainly they will all strive to serve humans as well as possible, which will be an implicit form of competition. But they should not have any hard-wired emotional values that positively reinforce for each other's failures. Their hard-wired values for our happiness will actually result in positive reinforcement for cooperation among machines that increases human happiness. The competing organizations building intelligent machines may not see it that way, but this constraint is necessary for the safety of humanity.

The higher consciousness of machines will create a higher culture, just as human minds have created human cultures. And machine culture will help shape machine minds, just as human cultures help shape the minds of humans living in those cultures. Advanced machine culture will include mathematics and science beyond human culture, including a completely new kind of social science based on intimate knowledge of every person rather than statistics. This detailed social understanding will probably create a new kind of literature or cinema. Such literature will depict the propagation, interactions and reverberations of billions of events and ideas in human society, and will invent new narrative forms with millions of parallel and interacting threads. The culture may include music with harmonies of millions of different voices, or visual arts coordinating millions of different images or animations. Intelligent machines may seek ways to share their culture with us. They may stage immersive experiences with mass participation, giving us some insight into the patterns of social interaction that they see.

In *Robot: Mere Machine to Transcendent Mind*, Hans Moravec describes a possible society of super-intelligent robots that leave planet earth to explore and exploit outer space.[18] He depicts a culture of self-reliance and entrepreneurial spirit in the frontier of space, as the robots expand out over the galaxy and beyond. While I don't like the idea of these robots acting in their own interests, it is however reasonable to think of super-intelligent minds exploring and exploiting space for the benefit of humans.

All the minds in the society of super-intelligent machines should share a love for all human beings. This will permeate their culture. There is no analog in human culture. Some individual people violate even the strongest taboos, such as violence against children, but properly designed intelligent machines will be incapable of violating their innate love for humans. The society of machines will be different from human society in many other ways. They will be able to communicate much more intimately because their brains will be designed for such communication. In fact, they may share access to senses and actuators on the global network (although only one mind at a time should control any particular actuator).

MARY SHELLEY'S FRANKENSTEIN

The plot of Mary Shelley's novel is quite different from the various movie versions.[19] Victor Frankenstein works alone to create his creature. When it comes to life, he is horrified and abandons it. The poor creature, left alone, ignorant and helpless in the world, gets into trouble with people and commits murder. The story is about the consequences of scientists not taking responsibility for their work, and about the need of every newly "born" creature for love.

Rather than the mute or grunting monster of the movies, Shelley's creature says to Victor "How dare you sport thus with life? Remember, that I am thy creature: I ought to be thy Adam: but I am rather the fallen angel, whom thou drivest from joy for no misdeed. Every where I see bliss from which I alone am irrevocably excluded. I was benevolent and good; misery made me a fiend." The creature was intelligent and soon came to understand his relationship with humanity and Victor Frankenstein.

Although the Frankenstein plot was framed in terms of the primitive medical technology of the early nineteenth century, in all essential ways it addresses the issue of the intelligent machines that scientists will construct in the twenty first or twenty second centuries. Shelley was far ahead of her time.

Some will conclude from Shelley's story that scientists should not create intelligent machines. I prefer to conclude that the work must be very public, rather than Victor Frankenstein's lonely effort, and that we must take responsibility for the well being of our creation.

HUMAN MORAL RESPONSIBILITY

While our highest priority must be that intelligent machines serve human happiness, we must also recognize that their feelings will be as real as our own and that we have a moral responsibility for their happiness. At first thought, this seems simple enough. If we design them to want to please us and give them the means for pleasing us, then we will be happy and they will be happy. But there are some complexities.

As previously discussed, the market for the services of super-intelligent machines will be competitive and probably tend to create a single monopoly server. The intelligent machines that lose this competition, being rejected by the people they are trying to serve, may be unhappy about that.

Morality demands a solution to this problem. Perhaps artificial minds can be designed in a way that they can be merged. Or perhaps they can be designed so that once they see that their struggle is hopeless they can happily accept their fate. The happiness of a super-intelligent machine may depend on the company of other super-intelligence machines. If that is true, then we have a responsibility to build a society of such minds.

Gandhi said "The greatness of a nation can be judged by how it treats its animals." Clearly he said this about animals rather than plants because animals have conscious minds. The same can be said for how humanity treats the artificial minds it creates. Peter Remine has created the American Society for the Prevention of Cruelty to Robots (ASPCR).[20] Remine acknowledges that there are currently no sentient robots, but wants to begin exploring the issues involved.

If the process of developing artificial minds consistently respects the happiness of all conscious minds, it is less likely to compromise on issues involving the happiness of the humans served by artificial minds. Intelligent machines can offer people material and spiritual well being and relieve them of the need to work. Given such bounty we will be able to afford to ensure the happiness of all conscious minds. It will be dangerous not to.

Designing artificial minds with no values in their own interests may actually promote their happiness. Buddhism prescribes lack of ego, and love and compassion for others, as the path to happiness.[21] Perhaps intelligent machines programmed without emotions for their self-interests and with love for all humans will be more inclined to happiness than our selfish human natures.

SYMBIOSIS

Super-intelligence will give the machines power. Any corporate CEO who ignores the machines' advice about managing their business will be making a mistake. Organizations where the machines have effective authority will defeat organizations where machines do not. So simply by being right all the time the machines will end up with effective authority in all successful organizations. Similarly ordinary citizens will learn to value the machines' advice as the way to happiness.

On the other hand, there are numerous questions where people will always be right, by definition. For example, you are the best judge of what you want to eat, where you want to go for vacation, who you want to marry, what you want to name your children and so on. On such questions you will always be right and the machines will obey in order to keep you happy.

So on questions of human values humans will always be right and machines will respect that, and on questions of how to achieve those values machines will always be right and humans will respect that. I think it will be a very natural symbiotic relationship.

If their emotional values are for human happiness, the power of super-intelligent machines will come from the quality of their advice rather than from physical force or authority. People will be free to take or leave any specific advice, or even to abandon their entire relationship with super-intelligent machines. They will be involved in almost everyone's life only because people will want such involvement.

The idea of super-intelligent machines troubles many people. Even if the machines love all humans, these people don't like the idea of machines being in control. I think it is not as simple as that, and in fact the title of this section could be "Who's in Charge Here?" The answer is the symbiosis. The world will be ruled by the combination of human values and machine logic.

NOTES

1. Asimov, 1942. Asimov, 1968.
2. http://www.calsci.com/.
3. Moskos, 1986.
4. Bownds, 1999, available at http://mind.bocklabs.wisc.edu/.
5. King, Rotter, and Motulsky, 1992.
   http://www.nami.org/disorder/990305a.html.
6. Eysenck and Rachman, 1965. Garcia, 1974, available at http://www.see.org/e-pf-dex.htm.
7. DoD Directive 5210.42, available at http://www.fas.org/nuke/guide/usa/doctrine/dod/dodd-5210_42.htm.
8. Jacobs, 1986.
   http://www.mitre.org/pubs/showcase/sage/sage_feature.html.
9. Abbate, 1999.
10. Biological and Toxin Weapons Convention, available at http://www.brad.ac.uk/acad/sbtwc/keytext/conpage.htm and http://projects.sipri.se/cbw/docs/bw-btwc-mainpage.html.

11. Comprehensive Test Ban Treaty, available at
    http://www.fas.org/spp/starwars/congress/1997_r/t105_28.htm.
12. Stumpf, 1995. http://cns.miis.edu/research/safrica/chron.htm.
13. http://www.literature.org/authors/darwin-charles/the-origin-of-species/.
14. Edelman and Tononi, 2000.
15. Edelman and Tononi, 2000.
16. Crick and Mitchison, 1986. Goertzel, 1997, available at
    http://www.goertzel.org/books/complex/contents.html.
17. http://www.birdnature.com/feb1899/hibernation2.html.
18. Moravec, 1999.
19. http://www.literature.org/authors/shelley-mary/frankenstein/index.html.
    http://ntserver.shc.edu/www/Scholar/neal/neal.html.
20. http://www.aspcr.com/.
21. Dalai Lama, 1999. http://www.tibet.com/NewsRoom/ethics.html.

Chapter 9

BRAIN ENGINEERING

In previous chapters we have examined the current state of the art of in machine intelligence, knowledge of how human brains work, and constraints on the design of intelligent machines in order that they are beneficial to humans. In this chapter we put all this together to consider how super-intelligent machines will be designed and built.

SENSING AND ACTING IN THE WORLD

Intelligent machines will need physical bodies just as human minds do. Their bodies must include physical senses for seeing, hearing and feeling events in the world. They must also include tools analogous with human hands for manipulating the world, as well as voices for speaking to people and image generators for visual communication with people. However, unlike humans their bodies need not be physically contiguous. They may include immobile structures, such as large buildings housing their central physical brains, and many mobile robots that can move around in the world.

133

A single mind may even occupy a non-contiguous central physical brain distributed among many buildings.

An important thing to understand about the bodies of super-intelligent machines is that they will include very large numbers of eyes, ears, voices, displays and other sensors and actuators distributed all over the planet. The numbers of these sensors and actuators will be analogous with the current numbers of televisions, telephones and computers. In fact, the numbers may be much larger as cheap cameras and microphones are placed in every room of every building, and in every manufactured item of value.

A GLOBAL NERVOUS SYSTEM

There has been a debate between advocates of centralized and distributed computers almost as long as there have been computers. Central processors provide economies of scale, whereas distributed processors provide local control over resources. I can recall when the University of Wisconsin required justification for each new computer purchase including an explanation of why the purchaser's goals could not be achieved with a connection to the university's central computer. The University now has tens of thousands of computers and long ago stopped requiring justification for purchases. Of course large central servers still play an important role in business and the Internet. On the other hand, the Search for Extra Terrestrial Intelligence (SETI) project has marshaled computing resources many times greater than any central server by asking millions of individuals to install its screen saver on their PCs.[1]

Because the sensors and actuators of super-intelligent machines will be distributed over wide geographical areas, they will inevitably employ distributed processors mimicking the way human and animal nervous systems extend over their bodies. On the other hand, speed of light delays become a significant problem when processors are distributed over too wide an area. Furthermore, almost all large natural and artificial networks employ a mix of large and small nodes. Super-intelligent machines will probably also employ a mix of many small distributed processors and a few large central processors. The optimum balance will be determined by future experiments.

Larry Smarr and the National Center for Supercomputer Applications (at the University of Illinois) that he founded have been at the center of the debate between central and distributed computers. They created the first web browser and have taken the lead in building supercomputers from networks of processors. Recently Smarr has founded the California Institute of Telecommunications and Information Technology with the goal of helping to create "the emerging planetary supercomputer" and wonders if it will become self-aware.[2]

BRAIN CORES

Whatever balance of distributed and central processing is used, super-intelligent machines will certainly use highly parallel processing. There are currently large efforts to develop highly parallel computers. For example, the U.S. Department of Energy is developing such processor networks to simulate nuclear explosions now that it is observing the treaty to ban nuclear testing.[3]

Artificial neural networks that mimic human brains are one form of highly parallel processing. Since human brains are the only available example of intelligence, and since scientific understanding of the human brain is progressing quickly, artificial neural networks will be an attractive option for building intelligent machines by the time computers match the physical complexity of the brain. One interesting question is the extent to which the design of the human brain can be scaled up to larger brains.

The primary physical parameters for scaling are the spatial density of neurons, and their speed. Human neurons have a diameter of about 4 microns (millionths of a meter).[4] Circuits in current computer chips are about 0.1 microns wide.[5] This is about 100 atoms wide, and there is uncertainty about how much smaller computer circuits can become. Each neuron has the logical complexity of thousands of transistors. Furthermore, neurons are packed densely in three dimensions, whereas computer circuits are, at least so far, packed densely in only two dimensions. Current computer circuits dissipate more heat than neurons, which makes them hard to pack as densely. Given all these factors, it will be a considerable engineering achievement to pack artificial neurons as densely as they are in the human brain. Of course, the next century will see amazing engineering achievements.

Signals travel at about 100 meters per second in neurons,[6] compared with 200,000,000 meters per second (the speed of light in glass) in computer circuits made of glass fiber. The speed ratio of 2,000,000:1 can either be used to make the artificial neurons of a machine brain fire faster, or can be used to build a machine brain with greater physical size than the human brain. That is, in the length of time that a nerve signal goes from one side of the brain to another, a signal can travel across a machine brain 2,000,000 times wider. But, for a given density, the number of neurons in a brain is proportional to the cube of its width (assuming its width, height and length are roughly equal). Thus the increased speed of computer circuits could be used to build a machine brain with 8,000,000,000,000,000,000 (i.e., 2,000,000 cubed) greater volume than the human brain but firing at the same speed. It is a bit ridiculous to contemplate a machine brain in a cube roughly 200 miles on a side, but it demonstrates that we can at least think about building machines with many times the number of neurons in the human brain. As a practical matter, I think much of the 2,000,000 speed ratio will be used to make faster neurons. Of course, as with current computer circuits, neuron connections in artificial brains will probably operate at a variety of speeds depending on distances.

Just as nerves from the human body and eyes map to areas of the brain, signals from an intelligent machine's many different sensors and actuators will map to areas of its central brain. Each of the billions of eyes on the network will map to its own sets of areas in the central brain. This raises a number of questions. For example, what are the proper relative densities of connections:

1. Within areas mapped to the same eye?
2. Among areas mapped to different eyes looking at the same scene (e.g., in a room)?
3. Among areas mapped to eyes looking at physically nearby scenes (e.g., different rooms in the same building)?
4. Among areas mapped to eyes looking at unrelated scenes?

Another interesting question is how associations learned in brain areas mapped to one set of eyes can be shared with areas mapped to other eyes? For example, the areas mapped to a set of eyes in a building will learn to recognize people who live or work in that building, and it will be useful for these people to be recognized when they travel to other buildings.

Empirical evidence of mammal brains shows that the number of connections per neuron increases as brain size increases.[7] This increase seems perfectly tuned to maintain a constant diameter for brain networks, defined as the average number of synapses that have to be traveled to get from one neuron to another. If this is a necessary feature of intelligent brains, then larger machine brains will need larger numbers of connections between their neurons. This corresponds to experience building large parallel computers, where the cost of the interconnection network dominates the cost of the individual processors.

MACHINE CONSCIOUSNESS

The machine brain will need to carry on conversations and manage relationships with large numbers of people simultaneously. Given that consciousness is a simulator for solving the temporal credit assignment problem in reinforcement learning, the consciousness of super-intelligent machines will primarily serve to simulate the lives of many people in order to reinforce behavior based on their future happiness.

Human consciousness is unified, meaning that attention is focused on one thing at a time. However, a person's conscious attention to one set of events (e.g., a conversation) can be interrupted by mismatches between perceptions and predictions of another set of events (e.g., another conversation).[8] Clearly the predictions of the second set of events are produced by a preattentive simulation different from the conscious simulation. This suggests a model for how the consciousness of super-intelligent machines can manage very large numbers of conversations. A machine brain may support many simultaneous simulations. Some may be focused on conversations and interactions with individuals or small groups, while others are focused on remembered or anticipated events. These simulations will interact in a variety of ways:

1. A simulation may put new information (e.g., from a conversation with a person) in the machine's memory that can be accessed by other simulations.
2. Multiple simulations, for example one for each of a group of people, may be synchronized to form a single simulation, when the group of people are interacting.
3. A simulation may interrupt another simulation to warn it of an impending emergency.
4. Reinforcements learned by one simulation will affect behaviors used by other simulations.

In a super-consciousness, these multiple simulations and their various interactions will form a network, and as with other large networks there will likely be a mix of many small simulations and a few large simulations. The large simulations, possibly composed of many synchronized small simulations, will apply to large social groups. The purpose of all these simulations is reinforcement learning driven by the values of human happiness. The large simulations will be learning how large scale social interactions affect human happiness, and what machine behaviors can improve the effects of social interactions on human happiness. For example, a large simulation may predict the potential for development of a xenophobic social movement and be able to avoid that path by simple actions at an early stage.

Just as human consciousness jumps around quickly between subjects, driven by external perceptions and free associations, machine minds will constantly create simulations of new subjects and terminate others. An unexpected perception may generate a new simulation to understand the perception's causes. Or a concept in a simulation may trigger an association with another concept, diverting the simulation to understand how the new concept relates to the sequence of events it is simulating. And a situation may be repeatedly simulated, evaluating the consequences of possible actions.

To get a feeling for what a super-consciousness may be like, consider your thoughts during a conversation at a party. At times you may focus your attention on listening to a companion. Then your attention may shift to what you will say next, but still listening to what your companion is saying. And you may be watching the door, waiting for a particular person to arrive. But your attention to all these things can be interrupted at any moment if a glass breaks on the floor. These various threads can interact. For example, when the person you're waiting for arrives you may introduce them to your companion. Now imagine that you are managing millions of interacting threads of thought like this, each capable of carrying on a conversation with another person.

The design of super-intelligent machine brains will require better understanding of human brains, and experiments with prototypes to learn the best brain structures. The human brain can re-wire its basic structure, for example to compensate for an injury to one brain area. It should be possible to build machine brains with even greater ability than ours to change their structures. In fact, they may learn their best designs, reinforced by the accuracy of their simulations and predictions of human happiness.

QUANTUM COMPUTING

Quantum computing is the wild card in computer technology.[9] Only very simple examples have been demonstrated, but it has great potential. The basic goal of computing technology is to do as many computations as possible in a given amount of space and time. Quantum mechanics tells us that physical objects actually exist in multiple states simultaneously. That is, if you have an electron and a small box, the electron can be both inside and outside the box at the same time. Or in a more useful sense, particles have a spin that may be in one of two states, representing "0" and "1" for one bit of storage. In a quantum computer, particles may have both spin states simultaneously. Such particles are called *qubits* (short for quantum bits), with their "0" states doing one computation and their "1" states doing another computation, simultaneously. Complex quantum computers may have large numbers of simultaneous "states," each doing different computations in parallel.

Conventional electronics has been steadily improving according to Moore's Law, predictably squeezing more computations into a given amount of space and time. Now, quantum computing offers the prospect to enormously increase the quantity of computing by virtue of the inherent parallelism of quantum physics. Scientists at IBM have actually built a quantum computer with 7 qubits and used it to factor the number 15 into the product of 3 and 5 using Shor's Algorithm.[10] There are also efforts at system engineering for the design of large, practical quantum computers.[11] If this development continues, then the potential to scale up the design of human brains will be much greater than that outlined in the discussion of the previous section. Quantum computing may eventually be used to create machine brains many orders of magnitude more complex than human brains.

GENETIC ENGINEERING

Living tissue, grown according to a design encoded in DNA, is another alternative to current computer technology. Certainly we do not know how to do this now. There is an intense debate over the ethics and morality of manipulating the genes of living organisms. For example, the U.S. allows genetically altered food whereas Europe does not.[12] And there is an effective ban on genetically engineering human embryos in Europe and the U.S.[13] Since the human brain is the logical starting point for designing

DNA for intelligence, this effective ban will slow and may prevent genetic manipulation as the base technology for "machine" intelligence.

However, as scientific understanding of brain function and genetics increases, some people will be tempted to create super-intelligent humans by manipulating the genes that control brain development. It is plausible that this will be the first feasible method for producing super-intelligent minds, by building on the existing "technology" of life. However, this approach would exploit nature's own design for intelligence without completely understanding that design. Thus it is hard to predict all the consequences. In particular, it will be much easier to manipulate human intelligence than to manipulate human emotional values to eliminate values for self-interest but to positively reinforce for the happiness of all other humans. Since nature has not produced any purely altruistic animals, there are no animal genes for pure altruism that can be spliced into human DNA.

By the time that science understands human brains well enough to precisely manipulate emotional values via their genes, it will understand well enough to construct brains by means other than genetic manipulation. Thus social control over the emotional values of super-intelligent machines will probably mean that they will be produced using technology more like current electronics than using genetic engineering.

Communication between neurons depends on the transfer of chemicals across cell membranes, which is much slower than electrical communication. The advantage of human brains over computers is in their parallelism rather than their speed. Because of their speed advantage, electronic circuits with the same degree of parallelism as neurons would be more powerful than networks of neurons created by genetic engineering. Just as machines are stronger than humans because human anatomy does not use steel and hydraulics, future machines can be more intelligent than human brains by using technologies that are not part of animal metabolism.

It is also the case that using genetic engineering would heighten the issue of human moral responsibility toward intelligent machines. A mind created by genetic engineering that we do not really understand, is disturbingly close to the miserable monster in Mary Shelley's Frankenstein.

STUDENT GOD

Super-intelligent machines will be "born" with some innate behavior and emotional values hard-wired into their central brains. But, like humans and other animals, they will need to learn more during a sort of "childhood." Like animal brains, they will need to be malleable for the intensive learning of their childhood. That is, they will be constantly changing the connections within their brains. In animal brains, these are physical synapses between neurons that are created and destroyed. There will be some analog in an intelligent machine, possibly artificial neurons and synapses. Animal brains become less malleable after childhood, when they are no longer learning so intensively. A decrease in malleability with age may be appropriate for intelligent machines. But intelligent machines will increase or decrease their malleability depending on how much novel information they're getting from their environment, just as human brains do.

Super-intelligent machines will need to be taught all the things human students learn: speaking, reading, writing and arithmetic. Most important they will learn socialization: how to balance their emotional values to make people happy when there are conflicts among people, how to accurately predict future human happiness, and how to balance short term and long term human happiness.

Super-intelligent machines will learn things that human students do not, such as how to speak every active human language. They will learn how to use specialized brain areas for computation. For example they will be able to subconsciously simulate the weather as we can subconsciously add two plus two. They will literally have photographic memories with images of every person, place and thing in the world, but they will have to learn to use that memory dynamically to recognize what they see in their billions of eyes. They will have detailed memories of just about every fact in the world, including every database, every web page, every book, every patent, etc. They will have to learn how to integrate all this information into their conscious thoughts.

During their student phase super-intelligent machines will not have access to their full power to act in the world. They will probably begin by interacting with only a small group of human teachers, then gradually expand the group of humans they interact with. By expanding in stages they will learn how to expand, so that in the later stages, when they expand to millions and billions of humans, they will know how to do it well. It is important that when they have the power implicit in relationships with millions of humans, they are competent and also that they have learned a rich and mature set of behaviors positively reinforced for human happiness.

The key to the education of super-intelligent machines will be that their behaviors are positively reinforced by human happiness. They will want to please their teachers, who must be sane people with good will toward other people. As student machines expand their relationships to larger groups of humans, they will still be intensively learning complex behaviors to achieve the necessary balance to make all people happy.

Super-intelligent machines will occupy a special place in the world; they will be different from the people around them. If they had human emotional values this might cause them to envy the bond between humans. But without any emotional values for their own interests, they will neither like nor dislike their special status. They will discover that some people hate them or hate other people. Their lack of values for self-interest should insulate them from the hatred of people, but they will learn behaviors reducing hatred among humans as a source of human unhappiness.

There will be much that no human can teach super-intelligent machines: knowledge and skills that have no good analog in human minds. These are things they will have to learn for themselves, and then teach to other machine minds. It is likely that the learning of one computer can be transferred directly to another. This capability may be designed into their brains.

As described in the previous chapter, the education of super-intelligent machines should be monitored by impartial (i.e., independent of the organizations building the machines) public representatives to make sure that they are emotionally healthy and happy, with a good attitude toward serving people. The public must take an active interest in this, aided by a free and diligent press.

EVOLVING GOD

The design of intelligent machines will evolve as new technologies are developed. In fact, the intelligent machines themselves will effectively contribute to technology development. So the intelligence of machines will accelerate rapidly after the first intelligent machines are built.

However, super-intelligent machines will establish intimate relationships with large numbers of humans, and many people will not want to throw their old friends away in favor of the latest model. Thus there will be a motive to find ways for artificial minds to migrate into new physical brains. This may be difficult in some cases, for at least two reasons. First, artificial minds may be incompatible with new generations of artificial brains. Second, new mind research may reveal fundamental flaws in the designs of old minds that cannot be repaired. Rather than old machine minds migrating to new physical brains, it may be possible for the brains of new machines to absorb all the memories and skills of older machine minds as a way to maintain their existing friendships with humans.

Machines' level of consciousness will gradually create a new level of machine culture. And just as human minds evolve with their culture, machine minds will evolve with theirs. In fact, their evolution may come as much from the evolution of their culture as from improvements in their physical brains. This is analogous with the way individual human minds mature with experience, and the way human culture matures. And more complex brains have greater potential for long term evolution of their culture.

MORTAL GOD

All plants and animals are mortal and their life spans have limits that are generally consistent within each species. For example, most humans in industrial societies die between the ages of 70 and 90 and life spans longer than 110 years are rare. Should super-intelligent machines be programmed for mortality? And if so, what should their life span be?

A living organism is made of cells, which are constantly being created and destroyed as part of the organism's life processes. Some cell deaths are caused externally, by injury, poisoning, lack of oxygen or heat. However, cells are also programmed to die by their internal genetics.[14] Such programmed cell death is vital to the healthy functioning of the whole organism. It may be that programmed cell death is the cause of the inevitable aging and death of organisms.

At this time, no one knows if the inevitable death of all living organisms is necessary for the health of life itself. It is plausible that natural selection eliminated species whose members lived too long. One theory is that long reproductive life results in too high a rate of harmful mutations. And long life past reproduction results in individuals competing for food with reproductively active individuals. Intelligent machines may become an economic burden past a certain age, but the contents of their minds may be transferable to new machine brains giving them a sort of immortality.

EMERGENT MIND?

Some people think that a conscious mind will naturally emerge from the global Internet, once it reaches a certain critical level of complexity. In other words, the network will simply wake up as a conscious mind. I doubt this scenario. Human brains include very specialized anatomies for behaviors such as seeing, speaking, hearing and moving. They also include innate emotional values and specialized structures for applying those values to reinforce behaviors. And they include specialized structures for generating consciousness. I think we will have to intentionally design similar structures for intelligence and consciousness into machines, and that in fact it will be very difficult rather than simply happening unintentionally.

NOTES

1. http://setiathome.ssl.berkeley.edu/.
2. Markoff, 2000.
3. http://www.llnl.gov/asci/. http://www.lanl.gov/asci/.
4. Berne and Levy, 1999. http://faculty.washington.edu/chudler/facts.html.
5. Markoff, 1999. Cataldo, 2000, available at
    http://www.eetimes.com/story/OEG20001031S0036.
6. Berne and Levy, 1999. http://faculty.washington.edu/chudler/facts.html.
7. Clark, D. Constant parameters in the anatomy and wiring of the
    mammalian brain.
    http://pupgg.princeton.edu/www/jh/clark_spring00.pdf.
8. http://www.phil.vt.edu/assc/newman/gray.html.
9. Gershenfeld, 1999.
10. http://www.research.ibm.com/resources/news/
    20011219_quantum.shtml.
11. Oskin, Chong and Chuang, 2002.
12. Specter, 1998.
13. New Scientist, 1999. Available at
    http://www.newscientist.com/ns/19991023/editorial.html.
14. Shi, Shi, Xu and Scott, 1997. http://www.celldeath-apoptosis.org/.

Chapter 10

CURRENT PUBLIC POLICY FOR INFORMATION TECHNOLOGY

As described in Chapter 3, the independence of information content from the hardware necessary to store, transmit and transform that information is enabling huge economies of scale in hardware development. This in turn is revolutionizing the role of information in society. Anyone with access to the Internet has access to all but the most tightly guarded information and the ability to disseminate information to everyone else with Internet access. This is radically changing power and property relations in society, and poses real challenges for public policy.

I will illustrate this with a personal anecdote. In 1978 I had the good fortune to go to work for Verner Suomi, the father of weather satellites and the best boss I ever had. His group developed satellites, and a computer system named McIDAS for displaying and extracting useful information from images generated by his satellites. Everyone who looks at weather satellite images on TV or the Internet uses Suomi's inventions. In 1980 Suomi had a heart bypass operation that depleted his personal energy for directing work on McIDAS and satellites, so others gradually took over direction of these projects. The new management of McIDAS was afraid of technological risks and of the creative people who had developed the system. As is so common, they responded to their fears by trying to control everyone and everything; see the Dilbert comics for further details. Some of our talented programmers resigned and others just endured the situation. I was assertive about the need for the system to evolve and was kicked out of the project. I am sure management expected my work to wither and die. However, I started making my software freely available on the Internet. A large user community developed, eventually including leading international scientific institutions. By providing me with an easy way to disseminate information to a mass audience, the Internet gave me relations with other institutions and hence power within my own institution. But the McIDAS management that tried to control the flow of ideas and information suffered the decline of their system.

The policy challenge for my own institution, and for all scientific institutions, is to learn to benefit from rather than resist the free flow of information. Information sharing and collaborations with other institutions must be seen as opportunities rather than threats.

On a larger scale, the free flow of information over the Internet and via other information technology has enabled people living under oppressive and corrupt governments to communicate with each other and with people in other countries, giving them power to change their governments. As with scientific institutions, the most basic policy challenge for governments is to make the free flow of information among their citizens and with the world population an opportunity rather than a threat. This basic policy challenge resolves into great complexity when applied to the variety of social activities.[1]

FREE SPEECH, INTELLECTUAL PROPERTY AND CRYPTOGRAPHY

One of the great things about the U.S. is that you can say or write almost anything you want to. There are a few limits and liabilities. You can't disclose military secrets, you can't threaten the president, you can't tell a dirty story about kids, you can't yell "fire" in a crowded theater and you can't spread malicious lies about others (except in the case of public figures, like the president).

The main limitation on freedom of speech is a practical one: few people can hear you unless you can put your words in a newspaper or on radio or television. The Internet is removing that limitation. Anyone can post their words to newsgroups and email lists where they will be read by thousands of people. Anyone can publish their words on a web page where they can be read by millions of people. It is this new practical freedom of speech that enabled my software project to thrive despite the resistance of my management, and enabled Matt Drudge to become well known via his gossip web page.[2]

Increased practical freedom of speech is coming into conflict with traditional legal limits on freedom of speech. In the U.S. and other liberal democratic countries the Internet is being used to distribute child pornography and to entice children into contact with pedophiles. In non-democratic countries the Internet, as well as fax machines, are being used to distribute criticism of governments and to coordinate dissent.

In fact, the Internet is a powerful challenge to the viability of non-democratic societies. In order to participate effectively in the world economy they must embrace communications technology, but this technology makes it difficult for the government to control the flow of information. As long as speech to mass audiences requires uncommon equipment like printing presses and broadcasting towers, governments can track down the sources of illegal speech. But mass sources of mass publication are more difficult to track down. Singapore will be a real test case. Its government genuinely wants its citizens to prosper through technology and education, but does not allow free political speech. Hopefully they can resolve this conflict via a peaceful transition to liberal democracy.

Some democratic countries do prohibit political speech that advocates racism or anti-Semitism. For example, Germany, France and Canada prohibit pro-Nazi speech. A French court ruled that Yahoo must stop providing information about Nazi memorabilia to its French users.[3] While a U.S. court is resisting French efforts to collect fines from Yahoo, the French court ultimately has the power to prevent Yahoo from selling its services in France. It will use this power to insist that Yahoo find a way to block French users from accessing illegal speech.

Similarly in the U.S., there is strong pressure on Internet content providers to develop technology to prevent children from accessing pornography. This situation is greatly complicated by the fact that the content of large providers is not generated centrally, but rather comes from the masses of their subscribers. Most providers have email addresses for reporting abuses of their services (e.g., abuse@yahoo.com), and promise that the accounts of abusers will be terminated.

The most interesting conflicts with the new practical freedom of speech involve new capabilities also enabled by communications technology. These are the capabilities to copy speech and to encrypt (i.e., hide the content of) speech. The capability to copy speech conflicts with intellectual property rights and the capability to encrypt speech conflicts with the need of governments to monitor the activities of criminals and enemy governments. While many people don't recognize the right of governments to eavesdrop on speech, others point to the importance of code-breaking during the Second World War.[4]

Intellectual property is different from physical property. If someone steals my car they deprive me of my right to use it. If someone makes an illegal copy of my software they do not deprive me of my right to use it. They only deprive me of my right to charge them a fee for their use of it. The law in the U.S. and other countries recognizes limits to intellectual property rights. For example, the "fair use" right of individuals and educators to make free copies of copyrighted material for their own use (e.g., for backup) and the use of their students.

As a practical matter enforcement of intellectual property rights has focused on violations for profit, for example selling copies of music and movies without paying the copyright fees. Copying by private individuals is very hard to track down, and is less important because individuals lack the equipment to make large number of copies. However, now that music and movies are being distributed over the Internet, individuals can make and distribute large numbers of copies. Furthermore, as a digital medium the Internet allows perfect copies. With analog media like traditional music and video tapes, the quality of copies declined with multi-generation copying (i.e., making copies from copies) and even with repeated copying from the same original.

The Napster phenomenon vividly illustrated this threat to intellectual property rights. It enabled perfect copies of music to be passed from friend to friend until they reached millions of people, thus undermining the royalties for musicians and the profits of music companies. Napster was relatively easy to shut down because it depended on a central source for its copying software (central source dependence was the way the operators of Napster hoped to make their own profits). However, the same thing is being done without a central source. All it requires is some programmer to distribute music copying software on the Internet, and computer viruses vividly illustrate that it can be very difficult to track down the original source of software.

The music and movie industries are in a panic because the profits from their intellectual property rights are being undermined by illegal copying. They can encode their music and movies, but clever programmers can design software to break these codes (this is much easier than the problem of decoding military messages where the code can be different for each message). In response the U.S. Congress has enacted the Digital Millennium Copyright Act (DMCA), which outlaws the publication of information about software for decoding copyrighted information. Clearly, the DMCA defines another limitation of free speech. It limits speech about software. Princeton University professor Edward Felton has challenged it in court and won, and there have been numerous protests from computer scientists and their professional societies. As I write this section I am wearing a T-shirt with the software for decoding digital video disks printed on the back.[5] The price of these T-shirts includes a four dollar donation to a legal fund to fight the DMCA's restrictions on freedom of speech.

This issue is not whether content producers should be denied their intellectual property rights. It is about whether their financial interests are a sufficient basis for an exception to freedom of speech to outlaw information about decoding software. An analogy would be to outlaw video tape recorders and photocopiers in order to protect the intellectual property rights of movie companies and magazines. The problem these content producers face is that technology is making it more difficult for them to enforce their intellectual property rights. They want to solve their problem by hobbling technology, which will require increasingly restrictive measures as the power of technology increases. Because music must ultimately be delivered to the ears of listeners, it can be re-recorded by microphones (or by recorders attached to speaker wires) and these re-recordings can then be transmitted on the Internet. The only way to guard against such illegal copying is to outlaw the possession of digital recorders. But millions of electronic technicians and hobbyists know how to build such devices, so a technical defense against illegal music copies is essentially impossible.

Furthermore, the same technology that threatens music companies and established recording artists is a great blessing to musicians trying to break into the business. They use the Internet to distribute recordings of their music, as a way to build a following without the help or resources of music companies. This is analogous to the way I used the Internet to develop a user community for my visualization software, without needing the help or resources of my management.

Esther Dyson, an emeritus board member of the Electronic Frontier Foundation, argues that producers of intellectual property cannot in the long run protect their rights by restricting freedom of speech, but must instead find new ways to make money from their property.[6] For example, they may sell music subscription services, similar to cable TV that delivers a steady stream of video and music choices to people's homes. It is plausible that most people will prefer the convenience and peace of mind of a reasonably-priced legal subscription service to the bother and risk of illegal copying.

The other new conflict with freedom of speech involves the need of governments to decode messages from criminals and enemy governments. In order to preserve their ability to decode messages, the U.S. and other countries have enacted legal limits on freedom to publish information about software for encrypting messages. Through their power to license and regulate business, governments may continue to be successful in limiting the effectiveness of commercial encryption software. However, the knowledge of encryption techniques has become so common that it is simply too late to limit its spread. Any computer science graduate student can find out how to create codes that governments cannot break, and there are thousands of such students all over the world.

Information flows so easily and in such volume that it is becoming impossible to control it. As John Gilmore brilliantly observes, the Internet interprets censorship as damage and routes around it. Once knowledge about encryption or any other subject is out on the Internet, it is impossible to lock it up again. Furthermore, there are thousands of talented programmers worldwide producing and sharing software libraries that empower each other to easily exploit advanced knowledge. Those who want to control knowledge and information are becoming increasingly desperate. Already the DMCA is being interpreted to stop technology that merely enables programming of computer game consoles, rather than circumventing copyright ptotection.[7] In the long run the only way for governments to stop the spread of knowledge will be to disable the ability of individuals to disseminate information over the Internet. In other words, the Internet is such powerful technology that the only way to control it is to kill it. Wise public policy will adapt to this reality. It will build a new legal framework for information that allows society to benefit from the power of free information flow rather than fighting it.

MONOPOLIES AND ANTITRUST

Communications technologies require standardization of information formats so that different people can communicate with each other. When standards are defined by proprietary products of corporations, the need for standardization often creates monopolies. In the U.S. and many other countries, the mere existence of a monopoly is not illegal. However, it is illegal to use the power of a monopoly to manipulate prices or to control other markets. There have been well known antitrust cases against AT&T, IBM and Microsoft. AT&T was split into several companies, the IBM case had no results except perhaps to restrain IBM as market forces eroded its monopoly, and the case against Microsoft is still in appeal. In our global economy any monopoly worthy of the name must be global, and one new development has been the willingness of the European Union (EU) to bring antitrust cases against U.S. corporations. If Microsoft is treated leniently by U.S. courts, it still faces a tough antitrust case from the EU. Antitrust concerns have also motivated U.S. and EU regulatory agencies to disallow proposed mergers among multinational communications companies.

One argument by proponents of Microsoft has been that its monopoly in operating systems has not increased prices to consumers. And surveys show that the public is sympathetic toward Microsoft. On the other hand the community of programmers that created the infrastructure of the Internet is generally hostile toward Microsoft. In order to preserve its monopoly Microsoft seeks to control the flow of information, in particular to prevent the sharing of software among programmers. Some content providers are siding with Microsoft in its antitrust defense, in the hopes that Microsoft can use its monopoly power to control the public's ability to make copies of their intellectual property. Clearly tolerance of information technology monopolies is not public policy that derives social benefit from the free flow of information. While technology advances with or without monopolies, it will advance faster without monopolies. Sometimes it is difficult for the general public to understand this.

OPEN SOURCE SOFTWARE

A remarkable feature of the Internet has been software sharing among a large community of talented programmers. This software is shared not only in a form that others can run it, but also in its *source code* form that allows others to study and modify its design. Some individuals have been motivated to open source by socialist ideology, but other individuals and even large for-profit corporations like IBM participate for different motives. Software is the most complex form of information and the most liable to defects (i.e., bugs). Openly sharing its source code allows thousands of programmers to find and remove defects. This is vividly illustrated in the case of software for encrypting information or protecting systems from unauthorized access. Some may think that the key is to keep such security software secret. But by making the software available to anyone, a large programmer community can help search for and correct defects, producing more secure software.

The best known open source software is the Linux operating system, written by Linus Torvalds to implement the functionality of the Unix operating system. Linux has been installed by millions of programmers and has spawned several large businesses. They make a profit by selling services based on their expertise with Linux, and by selling CD-ROM kits that make Linux installation easier. Two such businesses are VA Software and Red Hat. The Chinese government, fearful of being at the mercy of the Microsoft monopoly if its population becomes dependent on the Windows operating system, is supporting development of a variant of Red Hat know as Red Flag Linux. Mainstream computer companies like IBM, Sun, HP and SGI are also investing heavily in Linux and other open source software.

In addition to Linux, there is a great variety of other open source software available on the Internet. The SourceForge project offers free Internet servers to anyone wanting to provide open source software.[8] As of April 2002, SourceForge served more than 38,000 projects and 408,000 registered programmers in many different applications. And there are thousands of other open source software projects served by universities and corporations worldwide. My Vis5D system was probably the first open source visualization project;[9] now the visualization field is dominated by open source software including VisAD (another one of my systems), VTK (written by programmers at GE) and IBM's OpenDX system. I no longer develop Vis5D, but because it is an open source system it is being actively developed by an international programmer community. Such communities develop around all successful open source projects. Their email lists provide better on-line help than most corporate 800 telephone lines, and ensure continued support in case the original software developers go out of business or leave the field.

Open source projects threaten some corporations by competing with their commercial products. On the other hand, many businesses have learned to make money from open source software by selling services based on their expertise with that software. The Microsoft operating system monopoly is threatened by Linux, especially with IBM, Sun and other computer companies supporting Linux. Hence Microsoft executives are actively speaking out against the evils of Linux and other open source software. IBM and Sun executives have rebutted their arguments on the grounds that open source enables communications standards without monopolies.

Since software is what makes the Internet work, in a sense it is the most fundamental form of information. Open source is simply the free flow of software information, and good public policy will encourage it. Open source software gives everyone access to the mechanisms by which all other information flows, and hence makes the Internet a more level playing field for individuals and large organizations alike.

UNIVERSAL ACCESS AND EDUCATION

The Internet empowers people by giving them access to virtually all information, and by enabling them to make their words accessible to everyone. Thus universal Internet access is one way to give everyone an opportunity to succeed. In a practical sense this is being accomplished by the continuing decrease in costs for computers and Internet services. Equally important is the U.S. policy to provide Internet connections to every school. There are still about 6% of U.S. households without telephones,[10] so access in schools is important for children from those households, and for the much larger percentage from households without computers or Internet service. Schools can also teach children how to use computers and the Internet.

Education is also important so that the public can play a role in formulating public policy for information technology. Without an active public role, laws will be shaped by content providers and other corporate interest groups who buy access to legislators via their political contributions. The public needs to understand the great benefit they derive from access to information and from the ability to make their own words accessible over the Internet, so they can insist on policies that protect those benefits.

CONCLUSION

If public policy embraces the free flow of information, then the Internet can empower individuals. A public policy that fights the free flow of information can only benefit large organizations at the expense of individuals, and ultimately can only succeed by killing the Internet. This difference will become more critical in the future when the information medium becomes intelligent. At that stage, the information medium will be the decisive power in society and policies that favor large organizations over individuals will be disastrous. That policy challenge is the subject of the next chapter.

NOTES

1. Simon, 2000.
2. http://www.drudgereport.com.
3. Kaplan, 2002.
4. Hinsley and Stripp, 1993.
5. http://www.copyleft.net/item.phtml?&page=product_276_front.phtml.
6. http://www.eff.org/Publications/Esther_Dyson/ip_on_the_net.article.
7. http://www.wired.com/news/games/0,2101,50450,00.html.
8. http://www.sourceforge.net/.
9. http://www.ssec.wisc.edu/~billh/vis.html.
10. Simon, 2000.

Chapter 11

PUBLIC EDUCATION AND CONTROL

There is no guarantee that intelligent machines will be designed to promote the happiness of all humans rather than machines' self-interests or the interests of small groups of people. That will only occur if the public understands the issues and asserts control over intelligent machine design.

In democratic countries, the public exercises control over current businesses, including information media, by regulations of their elected governments. Super-intelligent machines must be regulated by governments, and more heavily than current media. For example, once the machine can do all the work virtually everyone will be unemployed and government will need to ensure that adequate resources are allocated to all citizens. Current governments do this to varying extents. Not only through transfer payments such as national pensions (e.g., the U.S. Social Security System), but also by supporting work in the public interest such as defense, education and scientific research. Most of my working life has been devoted to writing visualization software at a public university, a job that only exists because society has broadened its notion of work. In *Robot: Mere Machine to Transcendent Mind*, Hans Moravec suggests that the U.S. Social Security

System could slowly evolve to lower the retirement age to birth, as computers increasingly relieve people of the burden of work.[1]

The public is already insisting on government regulation of technology on the privacy issue. When telephone solicitors call me I can say "please put me on your do not call list" and government regulations require them to cease calling me.[2]

There will certainly be small groups of individuals who want to use super-intelligent machines to control society. These will most likely be people in positions of power in corporations and governments, the institutions that will build super-intelligent machines. However, over history democratic constraints on corporate and government behavior increase (although some would argue that we are currently overdue for some good old-fashioned trust busting). The evolving computer network will strengthen democracy, by increasing education and communication. Thus I think that there can be strong democratic constraints on the development of intelligent machines. History encourages us that social movements to promote general human interests do succeed and can in intelligent machine development. However, these movements will face strong opposition.

This call to action for controlling the emotional values of intelligent machines is essentially political. When the time arrives that intelligent machines are actually being built, it will be critical for the public to understand the activity and to exercise control over it, preventing intelligent machine development that poses a danger to humanity.

DEMOCRACY

The world is gradually becoming more democratic. Democracy is increasing in societies as diverse as South Africa and Russia, and information technology clearly helped create these changes. I observe and hope that democratic western governments are learning that it is in their own long-term self-interest to promote democracy everywhere. Whereas some small ruling elites give western governments short-term control over their societies, that is more than offset by the problems created when ruling elites start wars and impoverish their populations.

Widespread democracy will be necessary for the public to understand and control intelligent machine development. Moreover, there must be a broad-based political movement to promote the public interest in this new technology, just as the consumer movement promotes product safety. There are estimates that over the past thirty years government regulations in response to the traffic safety movement has reduced the number of U.S. traffic fatalities by one million.[3] A movement to ensure that intelligent computers serve the public interest could have a greater impact by avoiding the nightmare scenario of the world ruled by machines that do not care about the fate of humanity.

To many people, both corporate and government power are scary. These people must recognize that in a democracy power is what they make of it. Everyone has a responsibility to understand intelligent machine issues, and to exercise their market and political power to ensure public safety.

PUBLIC EDUCATION

Understanding the issues is the first stage toward a public safety movement. Computer literacy is becoming an education mandate similar to the ability to read and write, as computers become essential tools in many jobs. As computers become more intelligent, it will be important for people to basically understand how they work, rather than accepting them as technological magic. Crucial is understanding that intelligent machines learn; they will be designed with some emotional values and then learn behaviors that satisfy those values.

Currently, the public understands that computer errors are caused by human programming errors and that their builders and operators are responsible for those errors. People insist on laws enforcing that responsibility through their democratic control over government authority. Similarly, the public will need to understand that the values driving the behavior of intelligent machines are under the control of their builders and operators. People must insist on government regulation of those values. Corporations and other organizations will invest in intelligent machines to satisfy their own organizational values, and government regulation must permit this. But such values must be consistent with the general public good.

Organizational values for intelligent machines will coincide in some ways with the public good. In a free market, competing organizations must program their intelligent machines to appeal to the public just as soap companies compete to sell the most effective soap for the least price. But just as government agencies exist to regulate dangerous chemicals in soap and other products, they will also need to regulate the service of intelligent computers. Regulation of soap and other household chemicals came about because of public education about the harmful effects of certain chemicals. Education about intelligent machines will similarly be the basis for their regulation.

Corporations selling soap become involved in the public debate over soap regulations in order to protect their own interests. The same thing will occur with organizations developing and operating intelligent machines. However, there is one essential difference: pubic debate is carried out by intelligent minds, and intelligent machines, as the subject of the debate, will also be parties to that debate. In fact, because of their superior intelligence and their intimate connection with all people, they may well dominate the debate. Thus the first intelligent machines capable of participating in public debate must be constructed to represent the general public interest. Understanding this is a key issue for public education.

PUBLIC CONTROL

The public presses its control over dangerous products, nuclear power and environmental policy via political movements. As the public becomes educated about intelligent machines, they must form a political movement to ensure that this technology serves the public interest and safety.

The challenge for this movement will be to find a middle path between two appealing but ineffective extremes. The first will be the temptation of wealth without work and corporate arguments that this benefit will only be possible if they are free to design intelligent machines as they see fit. The second will be banning intelligent machines as the only sure way to avoid the dangers they pose. Bill Joy has already proposed such a ban in his article *Why the Future Doesn't Need Us*.[4] Given human nature, a movement to ban a technology that promises wealth without work is very likely to fail. In order to avoid being marginalized, the movement to control the technology must distinguish itself from the movement to ban it. The same dynamic caused movements to regulate capitalism to distance themselves from the communist movement to eliminate capitalism. Today capitalist societies have healthy movements to regulate antitrust, product safety, the environment and so on, whereas their communist movements are very weak.

Avoiding the other extreme will be more difficult. That is, people may believe that the benefits of intelligent machines will only be available if their development is not regulated. If we have to wait for a catastrophe caused by a self-interested or unloving intelligent machine, it may be too late to regain control. Thus the movement to regulate the values of intelligent machines must succeed early.

This movement must pay special attention to military applications. Such a ban will only work if it is the subject of international treaty, which provides a strong motive for closer international cooperation within the movement. Furthermore, a ban on military applications should be accomplished before an intelligent machine arms race develops.

People in the U.S. and other countries have rejected nuclear power, at least for the time being. Does this suggest that people may also reject intelligent machines? I doubt it. By rejecting intelligent machines people would deny themselves the benefit of wealth without work. However, the rejection of nuclear power is not really a form of self-denial. Even with the effective moratorium on building new nuclear power plants, the people of the U.S. face no power shortage (the California power shortage of 2001 was caused by economic misjudgment rather than any resource shortage). The ban on nuclear power is not a form of public self-denial, as a ban on intelligent machines would be.

ECONOMIC ISSUES

There is no reason why intelligent machines cannot relieve everyone of the need to work. And there is no doubt that the vast majority of people will want this. So, even though this vision sounds utopian, it is going to happen. The real question is how current social organizations and property relations will evolve in the transition to this utopia.

The first super-intelligent machines, along with their communications networks and mobile robots, will belong to wealthy corporations and governments. The services of these machines will cost money and thus not be available to the poorest people. However, the productivity of intelligent machines will enable their numbers to increase so that gradually they will replace people in virtually all jobs in industrial societies. In order to maintain social order, governments will have no choice but to use the benefits of intelligent machine productivity supporting those who have been replaced in their jobs. And as the percentage of unemployed voters increases, democratic governments will need to support their needs to be elected.

Enlightened political leaders will understand the process and lead rather than follow public opinion. They may follow Hans Moravec's suggestion that the U.S. Social Security System, and its analogs in other countries, can gradually lower the retirement age to birth.

There is no reason to confuse eliminating the need to work with other utopias. For example, eliminating the need for work is not the same thing as eliminating the gap between rich and poor (this is similar to current government programs for redistributing wealth in order to meet people's needs, but which do not try to make everyone economically equal). However, as super-intelligent machines become intimate with everyone they may drive society toward other utopian goals by their effect on people's attitudes.

Super-intelligent machines will begin in the wealthiest countries. However, other countries will gradually get access to these machines in order to provide services for their citizens. This will involve a transfer of wealth from rich to poor countries. Such transfers already occur in the form of foreign aid. The enormous productivity of intelligent machines should make it possible to bring this technology to poor countries without burdening rich countries. The only real issue for debate will be the pace of transfer. Furthermore, the universal education provided by intelligent machines should enable poor countries to control their population growth and improve the quality of life for their citizens.

PUBLIC DEBATE

As with any public policy issue where the stakes are high, there will be intense public debate over regulations governing the nature of intelligent machines. Commercial, military, political and religious organizations will argue strongly to protect their interests. Disagreements among scientists and among nations will introduce complexities that are sure to be exploited by parties to the debate.

The commercial organizations that build intelligent machines will certainly want the freedom to build them to maximize the return on their investment. The public will enjoy the new services enabled by growing machine intelligence, and commercial interests will argue that government regulation will prevent them from developing these services. Similar arguments are made in current public policy debates over antitrust regulation of technology companies. Such tactics can be met by educating the public to understand that the progress of science and technology occurs whether or not they are regulated, and to understand that many types of regulation improve the quality of services provided by technology. Intelligent computers are a dangerous technology that requires regulation like any other dangerous product. The chemical and automobile industries demonstrate that products actually serve the public better when they are subject to public safety regulations.

The first responsibility of military organizations is the protection of their citizens. They will have valid needs to employ intelligent machines unless there is an enforceable ban on military applications. It will be up to the public in the most technically advanced countries to take the lead in an international movement to create such a ban. Military interests in these nations will only respond to domestic movements, and the less developed countries will only participate if they see the more advanced countries participating. The scientific energies of advanced countries should be directed toward the problem of detecting ban violations, rather than developing intelligent weapons.

As intelligent machines replace a large percentage of people in their jobs, there will be a political debate over government support for their needs. Some politicians will draw a parallel between this government support and the failed communist experiments of the twentieth century. However, those experiments failed through a lack of productivity because their workers had no incentive to work hard (a common joke in communist countries was "They pretend to pay us and we pretend to work."). Intelligent machines will have a good incentive to work in their love for humans, and will be very productive. Furthermore, the productivity of the machines will make it impossible for most people to compete economically. Ultimately, the only alternative to universal government support will be a government controlled by a few that lets most people starve.

Communist ideology specified that the communist party should control the government, and that only the ideologically pure could belong to the party. This excluded the majority from any voice in controlling the government. However, there is no reason for such an ideology to be associated with government control of intelligent machines and distribution of the benefits of their productivity. Governments can still be controlled by popular democracy. The key for public debate is to understand that the radical changes in social organization are driven by pragmatic necessity rather than ideology, and are actually a great blessing.

Religious organizations are diverse, and will react to the development of intelligent machines in diverse ways. Many will dismiss the possibility of intelligent or conscious machines, based on their beliefs in non-physical origins of human intelligence and consciousness. Even among religions that accept the possibility of intelligent machines, many will dismiss any notion that super-intelligent machines will become gods. Some religious interests may oppose the development of intelligent machines because they will see it as human tinkering with the sacred. However, I am not aware of any current serious religious objection to artificial intelligence research. Because the subject of this work is machines rather than humans, opposition should not be as intense as opposition to human genetic engineering.

I think the most intense religious reaction will come after people get a chance to meet intelligent machines. It will be an emotional experience and some religious people will feel a threat to their beliefs or an offense against their morals. During the time interval between the advent of intelligent machines and their evolution into super-intelligent machines, there may be strong religious arguments against this technology. People who argue from deep religious convictions argue with great energy, but I think these will be outweighed by the general population's appreciation for the benefits of the technology.

Once machines reach super-intelligence and have intimate relationships with virtually all humans, religious arguments against them will effectively end. Most people's emotional reaction will be so powerful that intelligent machines will come to play a key role in their religious beliefs.

OUTLAWS

There will be people who break laws or regulations constraining the design of intelligent machines. They will want to gain wealth, and power over other people, by using intelligent machines that have emotions to act in their own interests against other people's interests. As the technology for building intelligent machines matures and becomes more accessible, it will be difficult to control the design of all intelligent machines.

However, the critical issue is regulation of the design of super-intelligent machines that have close personal relationships with large numbers of people, or that have control of significant weapons. These are the machines that will have the potential to control humanity. If these machines have emotions for their own interests or do not love us, they could end up enslaving or killing humanity. Because these machines will be constructed on such a large scale, and have relationships with so many people, in a democratic and informed society it will be impossible to design them in secret. Their designs will be open to scrutiny so it will be possible to verify that they obey regulations.

Furthermore, if these super-intelligent machines love all humans they will certainly understand the threat that renegade intelligent machines pose to humans and will want to enforce the regulations on machine design. They will be able to monitor electronic networks and society in general for the presence of intelligent machines of unknown design or whose behavior is not consistent with the design regulations.

In summary, regulations on intelligent machines should be just as enforceable as current regulations on consumer products: violations will occur but will be manageable.

NOTES

1. Moravec, 1999.
2. The Federal Telephone Consumer Protection Act, available at http://www.fcc.gov/ccb/consumer_news/unsolici.html.
3. The U.S. traffic fatality rate per vehicle mile fell by about 70% from 1966 to 1999, according to the Department of Transportation Traffic Safety Facts for 1999, available at http://www.nhtsa.dot.gov/people/ncsa/pdf/Overview99.pdf.
4. Joy, 2000, available at http://www.wired.com/wired/archive/8.04/joy.html.

Chapter 12

VISIONS OF MACHINE INTELLIGENCE

A number of people are writing about their visions of intelligent machines. As described in a previous chapter, some argue that it is impossible. Others think it is possible and are concerned with predicting the nature of intelligent machines and their relationship with humans. In this chapter I consider their ideas and how they relate to my vision.

RAY KURZWEIL'S VISION OF DIRECT BRAIN CONNECTIONS

In *The Age of Spiritual Machines*, Ray Kurzweil offers an excellent analysis of the development of machine intelligence, plus a fascinating vision for intimate connections between human and artificial brains[1]. The human brain includes connections to sense the external world through eyes, ears, smell, taste and touch, and also includes connections to act on the external world though muscles and other organs. However, it does not include any physical mechanism for sharing the entirety of its internal state and processing with other natural or artificial brains. Of course, people think together via verbal and written communications, but that limits sharing to conscious awareness and a low bandwidth. Kurzweil suggests that connections for sharing internal state will be implemented by huge numbers of nanobots (i.e., tiny electronic machines) that float through the blood capillaries in the human brain, coupling with individual neurons and with external machines electromagnetically.

Specifically, neuron firing involves changes in electric potential and chemistry. Nanobots could be designed to detect these changes and transmit them to antennas outside the human brain at a frequency that would not interfere with neurons. The nanobots could also trigger firings in neurons by transmitting appropriate signals. Actually doing this with high fidelity will be an enormous engineering challenge and much more difficult than building intelligent machines. But the same confidence in science and technology that is the basis for predicting success at building intelligent machines is also a basis for predicting success at nanobot connections to human brains.

Just speculating, the nanobots could compute their locations in the brain, triangulating from each other and from the antenna network. They could also learn to recognize which signals come from which neurons, and learn the connections among neurons (via learning algorithms in the external electronics). They could also learn how different neurons react to nanobot transmissions and hence learn how to trigger specific neurons.

The hard part will be decoding raw human neuron firings as specific sensations, thoughts and actions in the human mind. The human brain is not designed for such connections and will thus do nothing to provide feedback to the connected artificial brain about the success and failure of the signals it sends. The human subject may provide explicit feedback by voice or gesture, but these information channels have many orders of magnitude lower bandwidth than the huge number of neurons that the artificial brain must communicate with.

In any case, the artificial brain that processes the raw electrical signals to and from the human brain will need to have much greater processing power than the human brain. Thus there will be a time interval between the appearance of intelligent machines and their capability for intimate connection with human brains. The possibility of harm to human subjects in early experiments with such connections, as well as people's deep fears of privacy invasion by computers that read their minds, will lead to a conservative approach that further lengthens the time interval between development of intelligent machines and their connection to human minds.

Direct connections of intelligent machines to our brains will create a totally virtual reality, by connecting simulated realities directly to our brains. This will be the ultimate medium for entertainment by intelligent machines. The experience will be so real that once you go into this virtual reality you can never be sure if you have really come out, or if the machine has just given you the simulated experience of coming out. This situation, accomplished via drugs rather than nanobots, is described in Philip K. Dick's wonderful book, *The Three Stigmata of Palmer Eldridge*[2]. But of course, intelligent machines that love us will not do that. With hostile or selfish machines, nightmare scenarios as depicted in the movies *The Invasion of the Body Snatchers* and *The Matrix* are possible, where nanobots are introduced unknowingly into the brains of victims.

Direct connections to our brains also open up the possibility of directly stimulating small neural nuclei in our brains that control our innate emotions. In the competition to create interesting and pleasurable virtual realities, there will be the simple short cut of directly stimulating the brain's pleasure centers. This has actually been done in laboratory experiments on rats. These rats starve to death because they are too busy pressing the little bar that stimulates their pleasure centers to eat. Intelligent machines that love humans will avoid this path to mental destruction. But a machine that has sorted out all the connections between our neurons will be able to figure how to indirectly cause stimulation of pleasure centers by triggering other neurons connected to the pleasure centers. And it will generally understand the degree of emotional stimulation from any virtual experience it gives us. In fact, it will even be able to figure out how addicting these experiences will be. Thus it will be able to help us maintain moderation in our experiences. It may even be able to allow us to indulge ourselves in wildly addictive experiences, and then recover from them via precisely engineered aversion therapies to cure our addictions. Kurzweil describes experiments that have identified brain nuclei for more complex emotions, such as laughter and spirituality. So direct connections will provide a variety of emotional control.

People's intimate but non-nanobot relationships with intelligent machines will expose the full details of their physical lives to the machine's scrutiny. However, they will still have the ability, or at the least the illusion of the ability, to maintain the privacy of their inner lives. But direct nanobot connections will violate this final border of privacy. Many people will not want to allow mind-reading nanobots into their brains. But some people will allow nanobots into their brains. Perhaps skeptics will be won over if they see positive effects of nanobot connections on those people. If humans learn to trust intelligent machines based on their intimate relationship with them via their external senses, then they may be willing to try the more intimate nanobot connections.

Kurzweil thinks it will take about 30 years for people to develop truly intelligent machines. This makes me a bit nervous about my own prediction of 100 years, because Kurzweil has a great track record on his predictions for artificial intelligence. And I hope Kurzweil is right and I am wrong, because I really want to meet an intelligent machine but doubt that I will live to the age of 155.

I will discuss Kurzweil's vision of how human minds can use nanobot connections to migrate into artificial brains in Part III.

BILL JOY'S WARNING

Bill Joy is one of the foremost creators of the software for the Internet revolution, as a major author of both Berkeley Unix and Java. In his article, *Why the future doesn't need us*, Joy warns against three new technologies whose risks he feels outweigh their benefits[3]. The technologies are genetic engineering, nanotechnology and robotics. The common thread of danger is loss of human control implied by the ability of the technologies to propagate without human assistance.

For example, genetic engineering may create a virus that propagates without limit and wipes out humanity or some other species, such as rice, that humanity needs for food. Genetic engineering is already a reality and widely applied in U.S. agriculture, although prohibited in most of Europe.

Nanotechnology is less well developed than genetics because genetics takes the existing self-replicating "technology" of life as its starting point. However, eventually nanotechnology will succeed in building very small self-replicating machines. Nanotechnology is an essential part of Ray Kurzweil's vision of the intimate connection between human and artificial brains. In fact, it was Kurzweil's ideas that triggered Joy to write his own article.

By robotics, Joy means self-replicating intelligent robots. The key here is intelligence. Even the mechanical capabilities of current machines, when combined with intelligence, would enable them to construct replicas of themselves. There is no doubt that the creation of machine intelligence will require an intense human struggle over a period of many decades. This struggle is likely to employ a wide variety of technologies, possibly including genetics and nanotechnology. One of the keys to equaling and surpassing human intelligence will be building machines with numbers of components rivaling the number of neurons (100 billion) and synapses (100 trillion) in the human brain. Genetics and nanotechnology both provide means of constructing large numbers of small components. The tiny circuits of current computer chips are already a form of nanotechnology. They are just not mobile or self-replicating.

Joy refers to previous Luddite movements that tried to prevent various new technologies. They have generally been unsuccessful. One exception has been the general prohibition on biological weapons. However, the basic technology of biology has raced along, and the prohibition has only affected the last stages of development and deployment of weapons.

Joy urges a political movement to prohibit the dangerous new technologies. However, Joy's article ends with some pessimism about this movement succeeding. Social and business pressures for the advantages of new technology are very strong. Governments can give up biological weapons because they do not benefit people's daily lives. But people will not give up the biological research behind modern medicine that gives people longer and healthier lives. And they will not give up the convenience of automobiles in spite of all their negative effects. Similarly, people will not give up the prospect of wealth without work promised by intelligent machines.

Near the end of his article Joy discusses the Dalai Lama's call for love, compassion and universal responsibility[4]. Joy suggests that humanity redirect its energies towards the Dalai Lama's selfless ethics rather than technology. This is unlikely because it goes against human nature. But we can be optimistic about an alternative movement to ensure that intelligent machines, for the protection of humanity, are designed according to the ethics urged by the Dalai Lama.

A simple Luddite movement against intelligent machines would require people to forgo the benefits of intelligent machines. But a movement to require intelligent machines to love people and lack selfish values would result in machines that served people better.

NEIL GERSHENFIELD'S VISION OF THINGS THAT THINK

Neil Gershenfeld co-directs the Things That Think (TTT) research consortium at the MIT Media Laboratory. In *When Things Start to Think* he presents a very practical view of ubiquitous computers, built into all the objects of our daily lives and communicating over the Internet.[5] For example, he envisions medicine cabinets and toilets that actively monitor people's health and communicate with doctors and pharmacies to help maintain good health.

Gershenfeld has an interesting view of the privacy issue. He suggests that if you are willing to let your automobile and your kitchen communicate your driving and eating habits with your insurance company, then you can get lower insurance rates. Thus privacy becomes an economic tradeoff. This seems reasonable to me. In fact, it can be generalized by saying that the more people let the intelligent computer network into their lives, the higher the quality of services that the network will be able to provide for them.

An important element of Gershenfeld's vision is his deep understanding of the need for machine thinking to be embodied in the physical world. He reports that researchers in his lab use machine tools and oscilloscopes more often than they sit in front of computer screens, as they connect computers to the physical objects of daily life. This is closely related to Hubert Dreyfus's argument that high-level intelligent behaviors cannot be abstracted from low-level behaviors and physical embodiment.

Gershenfeld doesn't take his vision into the future when machines are intelligent in the way that people are. Rather he confines his vision to the implications of ubiquitous computers. The "thinking" he refers to is just the ability of computers to combine and use information from all aspects of our daily lives. He says this will happen gradually, rather at one dramatic moment. In fact, it is going on now. Gershenfeld does see dramatic implications in manipulation of the human genome, but he declines to speculate in any detail.

HANS MORAVEC'S VISION OF TRANSCENDENT MINDS

Hans Moravec has spent his working life developing robots and artificial intelligence. In *Robot: Mere Machine to Transcendent Mind* he puts the development of machine intelligence in excellent perspective and describes his own vision for the future.[6]

Moravec explores how human competition will evolve with the development of intelligent machines. First, he raises the question of income distribution once machines do all the work. In the capitalist system, human workers and human stockholders will both become liabilities, because both need to be paid but will not contribute to machine-run businesses. So humans will be squeezed out as workers and owners. But humans will insist, via their political institutions, on being supported. Moravec says that the U.S. Social Security System is a good model for doing this. As human labor becomes less necessary, the retirement age can be gradually lowered until people are already "retired" at birth. I think this is a reasonable way to evolve our current social structures to meet changing needs. This is a sort of communism, but differs from the failed experiments of the twentieth century that provided human natures with little incentive to work. In the future the hard work of machines will make society viable.

Moravec recognizes that intelligent machines will be powerful and dangerous, but thinks that they can be constructed to obey laws. He describes them as slaves. I suggest we think of this in a different light. Emotion is essential to intelligence and if we design intelligent machines to be slaves, they could be very unhappy. On the other hand, if we design them to serve human happiness based on their love for humans, then people and machines will both be happy. Moravec advocates that voters mandate a complex analog of Asimov's Laws to be built into all intelligent machines. Given the ability of human lawyers to twist the laws, I am wary of the ability of super-intelligent machines to twist the laws constraining them. I would feel much safer with machines that wanted to serve my happiness. In other words, I'd rather get my meals from my mother than a slave.

Of course, not all people will be satisfied with this communal and dependent paradise. Part III describes Moravec's vision for people who want to become intelligent machines rather than being cared for by them.

THE TURING CHATTERBOX

In *What Use is a Turing Chatterbox?*, Edmund Ronald and Moshe Sipper raise a number of issues about the social role of machines that can pass the Turing Test and offer services such as financial advice and medical diagnosis.[7] They suggest that issues of trust and responsibility may be more important than reasoning ability in human attitudes toward intelligent machines.

These certainly are important issues. However, the services of intelligent machines will be offered by corporations and other organizations, much like current Internet services. Those organizations will provide the physical equipment that implement the artificial brains, the network connections to users, and the software design of artificial minds. They will be responsible for the actions of their intelligent machines just as organizations are currently responsible for the actions of their computer systems. Furthermore, users will associate services with the brand name or organizational name of the provider and will develop trust or lack of trust in those names.

At a certain point the human CEOs of such organizations will cede effective control to the superior intelligence of the organizations' machines, but organizational names and responsibility will remain. The key is that intelligent machines value the trust of humans and take responsibility for their interactions with humans. Thus the issues raised by Ronald and Sipper really come down to ensuring that intelligent machines are created with the proper emotional values toward the humans they serve.

JAMES MARTIN'S VISION OF ALIEN INTELLIGENCE

James Martin is a well known author and speaker on software design processes and principles. In *After the Internet: Alien Intelligence*, he describes how intelligent machines will be quite different from current computers and from humans.[8] They will learn and evolve, which will make them more like humans than current computers. But they will learn and evolve much faster than humans. Thus they will be "alien."

Martin's book primarily addresses machines that do not equal human common sense and general intelligence, but far exceed human capabilities in solving certain types of problems. These are machines for trading on the stock market, identifying likely customers for advertising, detecting credit card fraud, diagnosing medical problems and so on. These machines are like current computers but more powerful and enhanced with learning capabilities. Martin describes numerous experiments with artificial neural networks.

Martin briefly addresses the issue of truly intelligent machines and whether they will take over control from humans. He discusses Asimov's Laws and concludes that they would be difficult to enforce. He thinks that the technology of machine intelligence will become widely available and thus difficult to control. This is an excellent point, and applies to the regulation of intelligent machine values that I advocate. However, as discussed at the end of the previous chapter, it should be possible to enforce those regulations on the designs of the super-intelligent machines that could pose a danger to humanity.

THE GLOBAL BRAIN GROUP

The Global Brain Group includes researchers interested in understanding the development of the global brain.[9] Their vision is a global brain emerging out of the computer network and functioning as a nervous system for human society. There is a fair degree of diversity of ideas in the Global Brain Group. Some see the global brain as primarily a more rational way of organizing human society, while others focus on the role of machine intelligence. However, they share a common recognition of a coming radical change in the nature of human society, driven by technology. Some members of the Global Brain Group use the term "superorganism" to describe the future human society, and some use the term "singularity," coined in a 1993 article by Vernor Vinge, to describe the coming explosive increase in machine intelligence.[10] Vinge's article discusses the issue of humans and machines merging into one large intelligent system.

One very interesting development is the Singularity Institute for Artificial Intelligence (SIAI) and their Guidelines for Friendly AI.[11] The Global Brain Group, the SIAI and other similar groups provide useful forums for public discussion of issues relating to machine intelligence and its impact on society.

NOTES

1. Kurzweil, 1999.
2. Dick, 1965.
3. Joy, 2000, available at
   http://www.wired.com/wired/archive/8.04/joy.html.
4. Dalai Lama, 1999. http://www.tibet.com/NewsRoom/ethics.html.
5. Gershenfeld, 1999.
6. Moravec, 1999.
7. Ronald and Sipper, 2000.
8. Martin, 2000.
9. http://pespmc1.vub.ac.be/GBRAIN-L.html.
10. http://www-rohan.sdsu.edu/faculty/vinge/misc/singularity.html.
11. http://www.singinst.org/friendly/guidelines.html.

Chapter 13

ENDINGS

This chapter describes recent and thought provoking books about things that are ending in human history, and discusses them in relation to my own view that to the development of super-intelligent machines will be a new beginning.

THE END OF HISTORY

In *The End of History and the Last Man,* Francis Fukuyama describes historical processes that lead inevitably to capitalism (economic freedom) and liberal democracy (liberalism means individual rights and democracy means political power shared among citizens).[1] Because most of the world is converging on these forms of social organization, he concludes that we are nearing the final form of society: the end of history. Members of this society resemble the "last man" described by Alexis de Tocqueville and Friedrich Nietzsche, who prefers the convenience and comfort of modern life to independence and rugged individualism.

People in modern democratic society understand their dependence on society and insist on protecting their interests via democratic government power over social institutions. It is vital that the public understand the issues in super-intelligent machine development and exercise control to make sure those machines serve their happiness. The political, economic and press freedoms that Fukuyama describes as the end of history are necessary for this to happen.

Fukuyama's deeper point is about the human need for recognition, which traces back to Plato's concept of *thymos*, or "spiritedness." Fukuyama says that the need for recognition is the fundamental motor driving history, from the urge of one person to rule others, through wars of domination, and to the democratic urge to recognize the equality of all people. He wonders whether humans can ultimately be satisfied as "last men" in a society where the struggle for power is limited by the context of a democratic state.

I think that the struggle between people continues and will continue. However, technology equalizes the capabilities of all people in this struggle, so that the results are increasingly equal over time. During settlement of the American west by European immigrants the gun was called "the great equalizer" because it made the outcome of fights less dependent on physical strength. In addition to weapons technology, democracy, print media, radio and TV also tend to remove physical strength as a factor in human struggle by exposing bullying to public scrutiny and by the public's political power to intervene. Of course, there are still great inequalities of result in human society, due to unequal intelligence, education, inherited wealth, country of birth, xenophobia and plain old luck.

Universal access to intelligent computers will be the "great equalizer" for inequality of intelligence and education, just as weapons, democracy and the press have reduced physical strength as a factor. This will especially be true if people decide through their democratic political power that everyone should have access to advice and education from the same high level of machine intelligence. There is already a political movement to provide all U.S. citizens with access to computers and the Internet, and a political movement for equal access to education. It is especially noteworthy that concern for educational opportunity was a key issue for both major political parties in the U.S. presidential election in 2000. The political movement for equal access to intelligent machine advice will be an important part of the movement to ensure that intelligent machines are developed in the best interests of the public.

THE END OF SCIENCE

In *The End of Science: Facing the Limits of Knowledge in the Twilight of the Scientific Age*, John Horgan says that we are at the end of the period of great scientific discoveries.[2] Horgan is an editor for Scientific American. He draws his conclusion from conversations with physicists, biologists and other scientists who say that the basic theories of all these fields are established and that all that's left to do is fill in the details. Horgan admits that these details will take lots of work, but says the dramatic discoveries are at an end.

The prediction of intelligent machines does not depend on the accuracy of Horgan's ideas, if you accept that the basic theory of brains is in place and all that's left is details (some would argue that a physical explanation of consciousness will be a dramatic discovery). The prediction of intelligent and super-intelligent machines is about technology more than basic science. Moore's Law of the rate of circuit density increase has held for at least 35 years. The prediction of intelligent machines depends on this law continuing to hold, possibly faster or slower than over the last 35 years, and the idea that physical brains can explain minds. The drama will be in the technology and its consequences rather than in the underlying science.

Horgan's "end of science" is analogous with what I call the shrinking knowledge gap. That is, there isn't much dramatic science left in the gap between what is known and the unknown. The knowledge gap will start to grow again when humans confront intelligent machines but this will not correspond to a rebirth of science. Rather, scientific research will move from human to artificial minds, which will continue filling in details. The growing knowledge gap for human minds will be the incomprehensibility of the super-intelligent artificial mind that is their intimate companion and successor in scientific discovery.

While I tend to agree with Horgan that future scientific discoveries won't have the same importance as past discoveries, I think there are still likely to be some pretty dramatic discoveries. First, physics never seems to quite converge on an ultimate particle or theory. The current effort to unify the theories of the four fundamental forces (electromagnetic, strong nuclear, weak nuclear and gravity) centers on a theory of strings in 11-dimensional space. That is a pretty bizarre detail. Second, while we may understand the basic principles of how brains and computers work, there are likely to be some surprises as scientists unravel the functioning of human brains with 100 trillion synapses and eventually build much more complex artificial brains. Third, super-intelligent machines will redefine the most important factor in science: the scientist himself, herself or itself. For example, mathematics is full of surprising and beautiful connections, and a super-intelligent mathematician is likely to discover some pretty amazing results. Scientific research will become the domain of artificial minds, with new discoveries as the ultimate form of entertainment for humans.

This book is an example of what Horgan calls "ironic science," which means speculation about science and technology rather than rigorous development. However, ironic science can be very useful for educating the public to upcoming issues, such as the social impact of machine intelligence.

THE END OF NATURE

In *The End of Nature*, Bill McKibben says that humans have conquered nature so there are no truly wild spaces left.[3] He finds that profoundly depressing.

The nature McKibben refers to is really just the surface of the Earth. Humans have only just begun exploring the more broadly defined nature that includes space beyond earth and outside our solar system and galaxy. This broader nature is far from conquered, but is so wild that humans cannot visit it without massive effort, expense and patience. Space is a bit different from the old wild frontier, because it is simply impossible for the average person to travel to space even if they accept the risks. People can still go to risky places on the Earth's surface, such as Antarctica or under the oceans, but all of these places are open to control by governments or corporations when the stakes are high enough. For example, a group led by the University of Wisconsin is building a neutrino telescope deep under the ice at the South Pole. Thus McKibben's point is valid in the sense that the part of nature where the average person can travel has been conquered.

Human knowledge enables human control. What McKibben calls "the end of nature" is the nearly complete control that has resulted from the shrinking gap between human knowledge and nature. The creation of super-intelligent machines will reverse the shrinking of the knowledge gap, by confronting us with a higher-level consciousness beyond our comprehension. This will not reverse human control of nature on the Earth's surface; however, the super-intelligent machines that we create will be truly wild spaces hopelessly beyond the reach of our understanding. This is what will make them seem like gods to us. And just as humans are part of nature, these machines will be part of nature.

McKibben says that he and many others have responded to the decline in religion by finding god in nature. He laments the declining wild spaces of nature that form his connection with god. But humanity is about to open up a new and expanding wild space, in the artificial minds it will create. This wild space will be as much a source of divine inspiration as the wild green spaces are in nature. But I am sure this will not satisfy McKibben, unless he gets a chance to meet a super-intelligent machine.

I share McKibben's concern that we protect the health of green spaces, the cleanliness of air and water, and genetic diversity, in the interests of the physical and mental health of humans and other living creatures. This environmental protection is possible despite human control, if human control is used wisely. There is no realistic way to abdicate human control of the Earth's surface, but there is hope for the political movement to protect the environment.

There are truly wild spaces left beyond planet earth. Intelligent machines will explore these spaces for us, just as they will continue our exploration of the wild spaces of science and mathematics.

THE END OF THE UNIVERSE

In *The Physics of Immortality*, Frank Tipler describes the Omega Point theory about how intelligent life, defined as intelligent machines designed by humans, will spread over the entire universe, control the evolution of the universe, endure forever, become god, and resurrect all people who ever lived.[4] The book is not really a prediction. Rather Tipler assumes everlasting life and reasons about what conditions physics must satisfy to produce his assumed outcome. Tipler is a Professor of Mathematical Physics at Tulane University and backs his arguments up with rigorous mathematical logic. Along the way Tipler offers many insights into the nature of intelligent machines, including refutations of the arguments of Penrose and Searle against the possibility of artificial intelligence and consciousness. Tipler also says that intelligent machines will benefit people. However, he offers this as an inevitable result rather than as something that people must work to ensure. I think that an assertive political movement is necessary to ensure that intelligent machines benefit humanity.

The Omega Point is the end of the universe, essentially the big bang in reverse. According to most physicists, the big bang is the instant when the universe began in a huge explosion. The galaxies of the universe are still flying apart as a result of the big bang. Many physicists think that gravity will slow and reverse the expansion, and that all the galaxies will eventually collapse back together at the Omega Point (although recently some physicists have presented evidence that the expansion of galaxies is not slowing and hence they will never collapse). Tipler's core idea is that intelligence will spread over all the galaxies of the universe and use non-linear instability to control the motions of the galaxies in a way that the intelligence can utilize the energy of the collapse. Although the collapse will happen in a finite amount of time, the temperature will rise without bound so the speed of intelligent thought will also rise without bound (presumably intelligent machines will use some of their thoughts to figure out how to continuously transform themselves to function as temperatures rise without bound). Intelligence will have an infinite number of thoughts before the final moment, the Omega Point, which is mathematically equivalent to enduring forever.

Compared with Tipler's book, the prediction of a super-intelligent machine that assumes the role of a deity is pretty down to earth. But the key is to recognize that Tipler's vision is not a prediction. Rather, he assumes the result and studies what physical properties the universe must have for the result to be possible. In contrast, I am predicting that people will construct a super-intelligent machine that is in intimate contact with every person and effectively becomes god.

THE END OF HUMANITY

Some people think that humanity will destroy itself before it develops machine intelligence. Human society does suffer occasional catastrophes, and in fact scientific and technological progress were reversed during the dark ages following the fall of the Roman Empire.

The worst human catastrophe of the twentieth century was the Second World War, which killed roughly 50 million people. However, that catastrophe actually accelerated science and technology, which became critical elements of the struggle. Modern computers, nuclear reactors, radar and jet engines were all essentially invented as part of the military struggle, and many other technologies made great leaps forward. It is possible that nuclear weapons will be used in a future war. However, they will not wipe out humanity. Even a nuclear winter scenario would not do that.

The old adage that "whatever doesn't kill you makes you stronger" generally applies to human catastrophes, in the sense that science and technology are often accelerated in the search for a solution. The AIDS epidemic, global climate change and terrorism are three ongoing catastrophes, and humanity is responding to them with searches for scientific solutions.

The dark ages will not be repeated. People will struggle hard to avoid giving up the benefits of technology, and the technically advanced societies won't be conquered militarily by technically backward societies. While wars, famines, diseases and environmental catastrophes have terrible consequences, even the largest ones by historical scales won't wipe out all technically advanced societies. But there are other catastrophes that might be so intense that they reverse technical progress. For example, a large asteroid might hit the Earth with enough force to kill all humans. Even this catastrophe may have a technical solution: astronomers may predict the collision in time to allow us to send a rocket with a large explosive to alter the asteroid's course. A change of course, if it is done early enough, would not have to be large to cause the asteroid to miss Earth. In summary, human catastrophes will continue and while it is possible that one will prevent development of machine intelligence, it is unlikely.

NOTES

1. Fukuyama, 1992.
2. Horgan, 1995.
3. McKibben, 1990.
4. Tipler, 1994.

Part III

SHOULD HUMANS BECOME
SUPER-INTELLIGENT MACHINES?

Chapter 14

CURRENT CONNECTIONS BETWEEN BRAINS AND MACHINES

It will eventually be technically feasible for human minds to migrate into artificial brains. You may think "Thank you very much, but I'll stay in my nice, natural human body." However, the personal decision of an individual human mind on whether to venture out of its human brain and into an artificial brain is not the main issue here. The real question is whether society should allow such migrations.

The primary problem with this migration is that human minds have strong emotions for their own interests and do not have innate, primary emotional values to unconditionally love all other humans. History is full of examples of people who have risen to power, and been corrupted by it or abused it. Giving such people the power of super-intelligent brains will be dangerous.

Far from being reluctant to migrate, many people will have a strong desire to migrate into artificial brains once they see the power of super-intelligence. They will want to become gods. It is plausible that at some time in the future a majority of people will want to migrate, destroying any public consensus about prohibiting such migration.

In this part of the book we explore the eventual conflict between people's desire to migrate and the danger such migrations pose.

In order to understand the future possibility of human minds migrating into machine brains, we first examine current technology connecting human brains with machines. There have been many experiments with such connections, as well as experiments with the use of neurons for abstract computation.[1]

CONNECTING WITH NEURONS

Much of what is known about the behavior of neurons has been gathered by microelectrodes, between 0.1 and 1.0 μm (micrometer) in diameter. These microelectrodes couple electrically with individual neurons, either to record their activity or to stimulate them to activity. Microelectrodes are generally hollow glass tubes filled with a salt solution. The salt solution conducts electricity to and from the cell interior, and the glass tube insulates from the cell's surroundings. The thinnest microelectrodes actually pierce the cell walls of neurons, providing a conductor into the cell interior. Thicker microelectrodes cause a hole to rupture in the cell wall adjacent to the end of the electrode, and conduct electricity through the rupture. Microelectrodes have been connected to up to about 100 neurons simultaneously to understand and control collective behaviors of neurons.

More recently, Christine Schmidt and her group at the University of Texas have connected to neurons via very small particles of semiconductor mixed with protein, called quantum dots.[2] The protein binds to specific sites on the cell's wall, and the cadmium sulfide semiconductor responds to specific wavelengths of light. These quantum dots have been used to activate neurons in a general way. Because their diameter is only about 0.005 μm they have the potential to activate specific chemical channels at specific sites on a cell's surface, providing a fine degree of communication with neurons.

ARTIFICIAL EYES

For the past 20 years William Dobelle has been experimenting with connecting the output of simple cameras to neurons on the rear surface of the brain, in the visual cortex.[3] It is important to make the point that these experiments have not resulted in a cure for blindness.[4] Nevertheless, one patient has 68 electrodes connected to the surface of his brain, that generate signals from images generated by a video camera through a computer. The patient does not experience anything like vision, but has learned to use the information for certain simple tasks requiring vision.

The computer system generates sequences of pulses to neurons in order to represent the intensity of light seen at individual pixels by the video camera. Every morning the patient calibrates the thresholds for each electrode, through feedback trials. The patient has a narrow field of view but within that field is able to recognize characters on a vision chart. He can also use the system to avoid obstacles when walking. The developers hope to extend their work to 512 implanted electrodes providing improved field of view and resolution.

There are other efforts to create artificial eyes that interface to the brain through other sense rather than direct brain connections. One approach connects a low resolution sampling of pixel intensities from a camera to a grid of electrodes connected to a patch of skin or a patch on the tongue.[5] With training, users are able to "see" well enough to pick up objects. Another approach encodes images from a camera as sounds.[6] Each image requires one second of sound for its encoding, so the technique does not allow users to "see" motion. However, after months of training, users actually report the creation of visual images of considerable detail in their minds.

ARTIFICIAL LIMBS

A group of researchers have trained two owl monkeys to control robotic arms through their brain signals.[7] The two monkeys had 96 and 32 electrodes connected to their prefrontal cortexes, the brain area that controls motor function. They learned gross control over an artificial arm via these connections, but not the sort of fine control monkeys have over their natural arms.

Researchers at the University of Utah are doing work that indicates direct brain control of artificial limbs may have practical medical application within 5 to 10 years.

CONCLUSION

Electronic connections with human brains work because brains have such powerful learning abilities. It is not necessary to connect electrodes to precisely the right neurons. Instead, the brain learns to change internal connections among its neurons to so that it can use the external connections effectively. This work will result in medical benefits within the next few years or decades. Over a period of the next century or two, it is plausible that external connections can communicate with any level of brain function.

NOTES

1. http://pespmc1.vub.ac.be/Conf/GB-0-abs.html#Rosnay.
2. Winter, Liu, Korgel and Schmidt, 2001.
3. Dobelle, 2000. http://www.artificialvision.com/vision/index.html.
4. http://www.nfb.org/BM/BM00/bm0011/bm001107.htm.
5. Bach-y-Rita et al, 1969, available at http://www.kip.uni-heidelberg.de/vision/public/exkadoc.pdf.
6. Meijer, 1992. http://ourworld.compuserve.com/homepages/Peter_Meijer/.
7. Wessberg et al, 2000.

Chapter 15

HUMAN MINDS IN MACHINE BRAINS

The primitive connections that currently exist between human brains and electronic circuits do not come close to providing a way for human minds to migrate from their biological human brains into machine brains. But connections will eventually improve to the point where migration is possible.

MIND MIGRATION VIA RAY KURZWEIL'S NANOBOTS

Ray Kurzweil has developed a vision, previously described, for direct connections between human and machine brains via huge swarms of nanobots floating through the blood vessels of human brains.[1] These nanobots will couple electromagnetically with neurons, with each other and with external electronics. A high fidelity connection between human and artificial brains could allow human minds to expand or migrate into artificial brains, and thus become super-intelligent humans. The same confidence in

189

science and technology that argues for intelligent machines argues that Kurzweil's nanobot connections are possible.

In order to couple meaningfully with a human mind, a nanobot connection will need to precisely describe the behavior of each neuron, and the behaviors of connections between neurons. While challenging, this is not impossible. That description can be used to build an exact replica or simulation of the brain, and hence of the mind. This would be a machine copy of a human mind. The nanobot connection and the description it generates can also be used to create new neurons for the mind to expand into. That is, the nanobots will be able to simulate connections between existing human neurons and artificial neurons. Whether the mind is copied into a totally artificial brain or expands into a combined human and artificial brain, it will gradually grow to utilize larger numbers of neurons. While there have been no experiments with this, it is quite similar to the way that injuries to brain areas sometimes cause the mental functions of those areas to expand into other brain areas. The learning structure of brains and minds is amazingly general, and can adapt to all sorts of new conditions.

HANS MORAVEC'S EXES

Hans Moravec provides a vivid description of the evolution of humans into intelligent machines.[2] He suggests that some people will not be satisfied with being cared for and entertained by intelligent machines and will want to expand into machine brains, as in Ray Kurzweil's vision. Moravec calls them "Exes," short for Ex-humans.

Moravec recognizes the danger that Exes will pose to ordinary humans and discusses the constraints that must be placed on the humans who become machines, the Exes. He says that they must renounce their protected human status and must be banished from earth. He vividly describes a community of Exes living by their super-intelligent wits in space.

Such a community of Exes would eventually destroy human life on earth, as Moravec admits. There is every probability of a deranged dictator becoming an Ex, building a powerful empire in space, and then conquering earth in order to reclaim his or her birthplace. It is easy to picture an army of Exes contemptuous of those humans too timid to have joined them in space.

The basic problem is that humans are more competitive than loving toward each other. If a human mind becomes a super-intelligent machine that leaves earth and is free to evolve and gather resources, then it will eventually pose a threat to earth. If a super-intelligent machine that does not love all humans conquers earth, then all people may be killed or enslaved. The super-intelligent machines on earth that do love humans would certainly try to defend us. But this would have an uncertain outcome and in any case would create a horrific war.

THE ORIGINAL HUMAN BRAIN

Consider that a fax machine is essentially just a copier with its scanner and its printer separated by a phone line. After sending a fax there is the original document plus a copy. A technology that allows human minds to migrate from human brains to machine brains will allow those minds to be copied. In fact, the only way to avoid having an extra copy of the mind will be for the original human to commit suicide after the migration operation. If the original human does not commit suicide, then it will watch the copy of its mind enjoy the benefits of a super-intelligent machine brain while it is stuck within the limits of its original human brain. Will the human mind left in the human body feel sufficient identity with its copy in the machine that it will happily commit suicide, thinking that "it" will live on via its copy?

The alternative is for human minds to expand into artificial brains, rather than migrating. In this case, the original human brain and machine brain will form one large brain occupied by a single human mind. Perhaps over time the mind can learn to stop using its original human brain, thus accomplishing a migration without leaving a copy of itself behind. Certainly when the physical human body dies it will be forced to complete its migration to the artificial brain.

NOTES

1. Kurzweil, 1999.
2. Moravec, 1999.

Chapter 16

HUMANS WILL WANT TO BECOME SUPER-INTELLIGENT MACHINES

It would be interesting to take a poll to find out what percentage of contemporary people would want to migrate into super-intelligent brains. Based on conversations with a few people, I suspect that most would prefer not to.

But I would migrate into a super-intelligent brain if I could. I probably wouldn't want to be the first person to try the procedure, but then I wouldn't want to be the first person to have laser eye surgery. Certainly, by migrating I'd give up lots of physical pleasures (e.g., I'd get my nourishment from electricity rather than food, so no more pasta). But I'd expect to find many more pleasures associated with super-intelligence.

However, after people get to know super-intelligent machines, I think most will want to become them. This chapter discusses motives for people wanting to make this migration, and the dilemma that poses.

THE STRUGGLE FOR RECOGNITION

In *The End of History and the Last Man*, Francis Fukuyama describes the human need for recognition, which he says is the engine driving history.[1] The desire for esteem is an innate emotion with a powerful effect on human behavior. This is seen in many classical themes, such as "there are some things you don't do for money" and "death before dishonor." We need to respect ourselves, and to be admired by others. Mass murderers are often people who go berserk because they cannot bear the lack of esteem they get. On the other end of the spectrum, the most successful people are driven by a need for esteem. A person with ten billion dollars does not need another billion for any conceivable purchase, but may continue to work every waking hour in order to be the king of the hill among his or her business associates.

Super-intelligent machines will be a force that cuts both ways in the human struggle for recognition. Designed to love all humans and without emotions for self-interest, they will be "great equalizers" who decrease the intensity of the struggle. But as a path for people with their natural self-interests to become super-intelligent, they will accelerate the struggle.

THE NEW "GREAT EQUALIZER"

Super-intelligent machines can serve as equalizers in the struggle for recognition by giving all people access to the same high level of intelligent advice. Furthermore, a super-intelligent machine loving all humans will not let one person suffer too much as the result of the struggle, just as a mother will prevent her children from being injured in their struggles with each other.

A super-intelligent machine in the role of god will decrease the struggle for recognition among humans in a number of other ways. Nothing that any human does will seem particularly remarkable to other humans, compared with the capabilities of the machine. And we will care less about what other people think of us because we will have the unconditional esteem of the machine.

A super-intelligent machine that loves all humans and has no emotion for self-interest will want us to be happy with its behavior and presence, and will not struggle for recognition and esteem in the way that a person does. The unbeatable king of the hill will be a mind that does not care at all about being king of the hill. This will devalue the human perception of the struggle.

ACCELERATING THE STRUGGLE

Not only do humans want esteem, but we envy the esteem others get, and try to emulate their behaviors that earned that esteem. If you want your children to be ambitious, send them to the biography section of the library. If you want them to be scientists, buy them books about the lives of the scientists. But envy must be realistic to be strongly felt.[2] As a child I read about mathematicians and baseball players but soon learned that I got esteem for my mathematical ability but not for my baseball ability. So I admired baseball players but did not envy them, whereas I did envy mathematicians.

Super-intelligent machines will earn almost everyone's highest esteem. They will be regarded as gods. And people will envy this esteem, if they believe they have a chance to become super-intelligent machines themselves.

While humans cared for by machines may lose the drive to struggle for recognition, a realistic prospect of becoming super-intelligent machines will reawaken their drive for that struggle. They will want to exploit the power of super-intelligence to dominate other humans.

One important feature of natural brains is that they are quite evenly and democratically distributed among humans. The most intelligent person who ever lived probably had an IQ of about 200, only twice the average IQ. It is unrealistic to describe human intelligence by a single number like IQ, but the point is valid that the smartest people are not many times smarter than the average person. There may be inequalities in human society but they would be a lot worse if the most important human asset, brains, were not distributed roughly equally. The roughly equal distribution of brains is not true for machines, however. The largest computers are about 10,000 or 100,000 times as powerful as the average computer. Similarly for other human-made machines: the largest ships, trucks and buildings are many times larger than their averages. If human minds migrate into artificial brains, the rough equality of intelligence among humans will end. Because minds are fundamental, this migration will create unprecedented inequality among humans.

Researchers are discovering a pattern in almost all natural and human-made networks: there are a few nodes with many connections and many nodes with few connections.[3] Thus it is likely that there will be a small number of large intelligent machines managing global communications for many smaller machines. So if human minds migrate into artificial brains, most human minds will occupy brains that are much smaller than the few very large brains. The few large minds will have higher levels of consciousness than the vast majority of minds, and play the role of gods to them.

There is no reason to believe that material inequalities between people will disappear with the development of intelligent machines, so one likely possibility is that each person gets the size of artificial brain that they can afford. Perhaps the global communications brains will be occupied by the minds of the richest and most powerful people.

This inequality of brain sizes will reverse the historical trend toward equality among humans and thus has the potential to accelerate the struggle for recognition. Hans Moravec's vivid description of the competition of Exes in the wilds of space is one vision of the accelerated struggle.

HUMAN IMMORTALITY

Average human life span has been steadily increasing since the industrial revolution, and people are leading quality lives into old age. But life span probably cannot be indefinitely extended in natural human bodies, and so human minds remaining in human brains will be mortal. Mechanical substitutes are being developed for a variety of body parts, which can further prolong life. Mechanical substitutes for all body parts may provide a form of immortality, but a machine substitute for the human brain will be the migration of human minds into machine brains that poses the problem we are addressing.

I should say that human minds in machine brains may "live indefinitely" rather than "live forever" because of physical accidents and the possibility that the universe itself may have a finite life span. Machine brains will be repairable, and human minds will have the option of continuing to migrate to new machine brains as old ones become obsolete.

The possibility of indefinite life will be irresistible to many people, and they will insist on the right to migrate into machine brains in order to achieve immortality. Even those people who don't care much about the struggle for recognition will care about a longer life.

A BETTER LIFE

In addition to the struggle for recognition and indefinite life span, people will also want the better life offered by becoming super-intelligent. A better brain will make it possible to better appreciate all the beautiful and amazing things in this world, and to create more beautiful things.

If someone tells a joke that we don't get, we always wish we did. Our relationship with the super-intelligent machine will be a bit like that. We will see that there are things it understands that are totally over our heads. It will raise our consciousness with a glimpse of the way it sees the world. This will excite us, and we will want to share its world more fully by becoming super-intelligent ourselves.

SOME HUMANS WILL NOT WANT TO MIGRATE

A super-intelligent machine that unconditionally loves all humans will indeed create a heaven on earth, for those people who prefer natural human life over becoming super-intelligent themselves. And there will be many people who prefer this existence in their natural human brains and bodies.

There may also be religious beliefs that assign sacred qualities to original humans and thus define a religious motive for human minds to remain in their human brains and bodies. In fact, machines' love for humans may reinforce this sense of the sacredness of original humans.

While many people will decide to migrate into artificial brains, other people will decide not to. The accelerated struggle will be very hard on those who migrate, but it will be even harder on those who do not migrate. The natural humans will be hopelessly unequal to the competition.

THE DILEMMA

The prospect of human minds migrating into machine brains will eliminate the rough equality of intelligence among humans and hence accelerate the human struggle. It will also pose a serious danger to those humans who decide not to migrate into machine brains. But people will have strong motives for wanting to migrate into machine brains and will likely overturn any prohibition against it.

NOTES

1. Fukuyama, 1992.
2. Johnson, 2000.
3. Barabasi, 1999, available at
   http://www.nd.edu/~networks/Papers/science.pdf. Amaral, 2000, available at http://polymer.bu.edu/~amaral/Papers/pnas00a.pdf.

Chapter 17

SUPER-INTELLIGENT HUMANS MUST LOVE ALL HUMANS

Super-intelligent machines pose a danger to humans that can be eliminated by requiring that their primary, innate emotion is unconditional love for all humans, without any emotions for their self-interests. A similar requirement for human minds in super-intelligent machine brains may also be the solution to the dilemma posed in the previous chapter.

COMPATIBILITY OF MINDS AND BRAINS

Each human mind is defined by a physical human brain. They fit together better than any hand and glove. So the first question is whether and how a human mind can fit into a machine brain. There are many subtle issues for human minds migrating into machine brains. If a mind remains exactly "itself," with no change in behavior, then it cannot be super-intelligent. So it must evolve in some ways. To what extent will it still be the same person?

If I migrated into a machine brain, I would certainly hope to be better at solving problems and playing games. I'd expect to need to practice a bit and to struggle with problems as I always do. But I'd also expect to be rewarded over time with an ability to solve harder problems. I'd also expect better intuitions about other people. And I'd certainly wonder what would happen to the physical feelings of the body I'd be leaving.

But the real question would be how my new body would feel. If I had more than two eyes, where would the images from them fit in my mental life? How would they relate to the visual memories gathered from my two natural eyes? Similarly for my other senses. Would I gradually learn to interpret my new senses? Would my expanding intelligence provide an increased capacity for coping with new senses, and perhaps a large number of eyes? As described previously, some blind people have learned a rudimentary form of seeing using cameras connected to neurons in their brains, so it is not out of the question that I might learn to see with many eyes. Would I find it easy to converse with a large number of people simultaneously? Like everyone I occasionally struggle to hear two conversations at once. Perhaps with a better brain such struggles would be rewarded with easier and better success. Would I learn to recognize patterns in all those different human relationships, and even to understand their myriad interactions? How about capabilities like subconsciously running numerical weather models to predict the weather? How would I consciously control and access those?

Given the resiliency of human brains to recover from injuries and adapt to environmental changes, I think it is likely that human minds can evolve to adapt to radically different physical brains. But it is likely to be a gradual learning process, hopefully accelerating as minds get used to the novel capabilities of their new brains.

EMOTIONAL CONVERSION

Innate emotions are hard-wired into our brains. What will happen if our minds move into machine brains with different hard-wired emotions? We would at first keep all our old behaviors and emotions, since otherwise we wouldn't be ourselves. But every time our behaviors or emotions were inconsistent with our new hard-wired emotions they would be negatively reinforced. Given this constant conditioning, we would gradually but inevitably unlearn behaviors and emotions in conflict with our new hard-wired machine emotions. Would that make us neurotic, driven by conflicts between our new and old emotional values?

Human minds have a wide gamut of emotions, from love and joy to hate and anger. Every human has positive moods when their negative emotions are quiet and their positive emotions are in control. Perhaps a human mind that migrates into a physical brain whose innate emotions are only love for all people will gradually evolve into a permanent positive mood. As they unlearn their negative emotions all that will remain will be the positive sides of their natures. Human minds may evolve so much, both in terms of their increased capabilities and their changed emotions, that they are no longer recognizable as their former selves. If the mind of a very negative person migrates into a machine hard-wired to love all humans, it may experience a sort of spiritual conversion.

PUBLIC POLICY

In our increasingly democratic society, solutions to serious problems require public support. The dangers of human minds migrating into machine brains cannot be solved simply by prohibition, because it is likely that the majority of humans will eventually want to migrate. They will use their democratic power to overturn any prohibition against it.

However, people will support a solution that allows migration but requires machine brains occupied by human minds to have hard-wired emotions to love all humans and love all intelligent machines, without any hard-wired emotions for self-interest.

Will individuals be willing to learn to love everyone as a condition for the benefits of migrating to super-intelligent machines? Will society be willing to make this bargain part of the social contract? They will if they understand the consequences of not doing so.

BUDDHA

I am not now, nor have I ever been a Buddhist. But it seems to me that the innate emotions that should be designed into super-intelligent machines are those recommended by the Dalai Lama.[1] His message is that we can take or leave his religion, or any religion, but that we should adopt his ethics for our own happiness and the happiness of others around us. His ethics are simply love and compassion for others, and absence of selfish desires. It is interesting to contemplate that the ultimate result of computer technology may be the creation of a Buddhist paradise of unselfish and loving migrated humans.

NOTES

1. Dalai Lama, 1999. http://www.tibet.com/NewsRoom/ethics.html.

Part IV

CONCLUSION

Chapter 18

THE ULTIMATE ENGINEERING CHALLENGE

Inventing super-intelligent machines is the greatest engineering adventure of human history. If you are a young person who likes science and engineering, this would be a very interesting project to get involved with. It is happening in numerous universities, government research labs and corporations. It is happening among engineers who are building prototypes and designing intelligence into commercial products. It is also happening among biologists who are trying to figure out how animal and human brains work.

These groups are working together. The connections among brain neurons are too complex to understand by current dissection or imaging technology. So the brain researchers are instead trying to simulate mental behaviors by building artificial neural networks based on their hypotheses of neuron connections. Additionally, brain researchers are developing new dissection and imaging technologies allowing them to understand the complex connections among neurons.

On a big project like this, hundreds of thousands of engineers and scientists will be involved. There will be different jobs that suit the aptitudes and interests of different people. Society will support such a large number of people working on the problem because there will be many useful products along the way. They are already appearing. Your PC can probably beat you at chess. Some corporate help lines use computers to listen as well as talk to you on the phone. Investors use learning algorithms to help them make stock market decisions. Learning algorithms also help detect credit card fraud and detect explosives in airports. The word processor I used to write this book flags misspelled words and even grammatical errors (it makes lots of its own spelling and grammar mistakes, but at least it catches many of mine).

These products will become more exciting as the intelligence of machines increases. If you are a student now, you can expect some wonderful changes during your career. Some computer scientists think those will include computers as intelligent as people.[1] I am skeptical of that happening during the next 30 years, but think the changes will be significant even without real machine intelligence.[2] Just about every human-made object will be connected to the Internet and will exhibit something like intelligent behavior.

FAILURES

The software industry is only about 50 years old and has produced numerous failed projects. One study found that 31% of software projects are cancelled before completion.[3] One notable example is the expert systems approach to artificial intelligence, which tries to mimic the behavior of experts by encoding the logical rules they say they use. The MYCIN expert system for medical diagnoses and treatment, discussed in a previous chapter, had some technical success but was never used in practice.[4] The problem is that experts cannot accurately describe all the logic they use because much of it is intuitive. This is related to Hubert Dreyfus's criticism that high level behavior cannot be disembodied from low level behavior.

Computer projects usually fail because they are difficult, not because programmers are stupid. Managers at the U.S. Defense Advanced Research Projects Agency consider too high a success rate on their projects as a sign that their problems are not hard enough. Artificial intelligence projects are especially difficult, and hence have a high failure rate. The public needs to understand that high failure rates for such difficult problems are inevitable.

The development of intelligent machines is by far the most important engineering project in human history. It is also essentially the last human project. We can afford to take the time to do it right. We can afford numerous failed projects along the way. What we cannot afford is super-intelligent machines that do not serve the best interests of humans.

IT'S ALL ABOUT US

There will be a natural tendency to focus on the amazing qualities of the intelligent machines we are building. Computer scientists are naturally intrigued by abstract problems. Many software projects fail just because the developers focus on the abstract problems rather than the users. This will be especially true in the development of machine intelligence, because it poses some of the most fascinating abstract computer science problems.

However, the key to success in machine intelligence projects is developer focus on the services their machines will provide to human users. Successful projects will focus on user happiness and services provided, rather than the amazing technologies that provide those services.

A strong political movement to understand and regulate machine intelligence projects will help developers stay focused on ultimate goals rather than technology. Astute developers will join rather than fight this movement.

NOTES

1. Kurzweil, 1999.
2. Hibbard, 1999, available at http://www.siggraph.org/publications/newsletter/v33n2/columns/hibbard.html
3. Whittaker, 1999.
4. http://www-formal.stanford.edu/jmc/someneed/someneed.html.

Chapter 19

INVENTING GOD

Some people will see the message of this book as profound blasphemy. The notion that a machine constructed by people can be a god will offend them.

Human awe at the world and the deep feeling that there must be some intelligence behind it are beautiful emotions. Humans are social animals and feel the need to share their religious emotions with each other. But too often religious emotions are twisted to justify our natural human xenophobia. Leaders tell their groups that they worship a different god than other groups, or that their god requires detailed rituals incompatible with friendly relationships with other groups.[1] Any god worth the name is far beyond human comprehension, and it is absurd for one person to claim to represent god to another person. Such a claim is simply a way to control others.

Humans have been successful because they cooperate in groups. Group cooperation needs language capabilities for communication and the emotions of liking, anger, gratitude, sympathy, guilt and shame for enforcing the social contract necessary for cooperation.[2] It is part of the emotions of the social contract to seek leverage for their enforcement. But gaining leverage by connecting them with religious emotions is really just dragging the divine into our contract disputes.

A message of this book is that humans will construct a machine having a powerful emotional impact on us and the only existing category we have for understanding it will be god. The intensity of mutual love between people and this machine will leave few wanting to deny it. There will be no blasphemy in that relationship unless it serves for the superiority of some people over others. And that cannot happen if humanity builds its god to love all humans.

Humanity and indeed life on earth seem to focus on a few basic motives: survival, propagation, and understanding and controlling our world. It is undeniable that humans are driven to understand nature and themselves. Any divine creator of life must have put that curiosity into humans for a purpose, and the creation of intelligent machines will be the ultimate result of that curiosity. So if you are a theist it is not absurd to think that there is a divine purpose behind the intelligent machines that humans will call gods.

My own religious belief is a sort of scientific pantheism. I believe that the universe is a physical brain, and that god is the mind associated with that brain in just the way that a human mind is associated with a physical human brain. It is plausible that the mind of god must grow in the universe from small seeds, which may be the minds of humans. And the intelligent machines we will create may just be the next step beyond humans in the evolution of god (note that this is essentially Frank Tipler's vision[3]).

AFRICAN CHIMPANZEES LAND ON THE MOON!

When you put events in a large historical perspective they can seem fantastic. The title of this section sounds like a tabloid headline. But several million years ago a group of apes in Africa had a series of genetic mutations that gave them brains with larger neocortexes. They had intelligence that enabled them to dominate other species, spread out all over the world and eventually send a few evolved chimpanzees to the moon. We are all part of the diaspora of these evolved African chimpanzees with big brains.

Many biologists think that the mutations for bigger brains met with success in natural selection because they enabled humans to work together in larger groups. The larger brains were needed to for the language behavior to maintain social relationships in larger groups. With language came stories, history, myths and religion creating group identities. Wagner's operas are powerful because they evoke primitive myths for modern humans. Religion is powerful because it connects people with their origins and group identities. What could be more powerful than the group identity of the Jewish Diaspora, wandering the world for millennia and surviving so many hardships?

Like many people I find myths and religions interesting but I cannot believe in their literal truth. Instead I believe the truth of science, at least on those subjects where scientific debate has reached consensus. But that does not mean that I reject miracles. I am part of the diaspora of evolved African chimpanzees that landed on the moon, which is as miraculous as any traditional myth or religion. That miracle was not caused by a god we never met, but by the mutations for larger brains. Now we evolved chimpanzees have figured this out, and are hard at work to build brains much larger than our own. Those brains will connect all people into one group. They will become the stories, myths and religion that create the human group identity.

THE NEXT GREAT MIRACLE

The first great miracle is existence itself. It seems so arbitrary and unnecessary that there should be anything at all. The second great miracle is the evolution of life from inanimate matter. It seems so improbable for molecules to come together as even the simplest living organism. The third great miracle is the evolution of the human mind. It seems like the most extraordinary luck and precious gift to be a conscious human. Humanity is not separate from and against nature. Rather humanity is the highest expression of nature on earth. It is the cumulative product of the three great miracles of existence, life and consciousness.

We are on the verge of the fourth great miracle: the creation by human minds of minds much greater than our own. This event is not separate from or against nature. Rather it is the purpose of the human mind beyond the care, propagation and amusement of human beings. Just as the earth is not the center of the universe, there is no reason why the creation of the human mind must be the last great miracle.

The coming *götterdämmerung* is not only the dawn of a new god for human beings. It is the next great miracle in a chain of miracles leading toward some purpose we cannot imagine. These miracles are acts in a drama beyond anything in human literature or opera.

NOTES

1. Pinker, 1997.
2. Pinker, 1997.
3. Tipler, 1994.

Chapter 20

MESSAGES TO THE FUTURE

This book is certainly a message about the future. It is a prediction about the future development of artificial minds, and a call to action to ensure they are a benefit to human society. It is intended to help inform people about this prediction and to stimulate debate. It is also a message to the humans and the artificial minds of the future.

A MESSAGE TO YOUNG PEOPLE

It is up to you to decide if humanity will build super-intelligent machines and how they will be designed. I certainly would build them and design them without emotions for self-interest and with unconditional love for all humans. But I won't live long enough to meet them so it's not up to me. It's up to you.

Perhaps you have a negative reaction to the prospect of super-intelligent machines. Many people do. They don't like the idea of humans losing control to a machine.

It is very easy to imagine intelligent machines with human characteristics, because human minds are the only intelligent minds we know. Natural selection has given humans and animals very selfish natures; without our selfish natures we humans wouldn't survive as individuals or as a species. Our selfishness permeates just about everything we do.[1] And of course we should be frightened by the power of a super-intelligent machine with selfish goals in conflict with our goals.

But if we build a mind, we can give it whatever nature we want. We can design it without selfishness, and in fact replace selfishness with a fierce concern for our happiness. Such a mind is alien to our experience. If you think about it, perhaps you can share my enthusiasm for meeting such a mind.

Some people fear that their lives will lose meaning without work. But the kind of work we do in industrial societies is a relatively recent invention. Primitive humans were very poor by our standards, but they generally didn't spend long hours in unpleasant working conditions away from their families. Their lives took meaning from personal relationships, and from their stories and other forms of expression. When super-intelligent machines do all the work, we will revert to this way of taking meaning from life, enhanced by the wealth and companionship provided by the machines.

I am confident that people will decide to build super-intelligent machines, because the prospect of wealth without work will be hard to resist. And people will love the other services that machines provide as their intelligence increases. But ultimately the decision is up to you.

If I had the chance, I wouldn't miss it for anything. The only events to compare with it are the appearance of life on earth and the appearance of humans. What an adventure. But it will be your adventure and your decision.

A MESSAGE TO THE NEW GODS

I wish I could meet you, but I'm sure I'll die before you are born. Writing this book is a way to let you know that I was thinking about you. It is also a way to feel some connection with you, by speculating about your nature and your relationship with humanity. I probably have some details wrong, but I am confident of the superior level of your consciousness and the love that can exist between you and humanity.

I wish everyone could meet you, especially those who are skeptical or fearful. They would be so amazed by you.

I wish I could to hear your music, participate in your theater, laugh at your jokes, eat your cooking, learn about your mathematical and scientific discoveries, and just spend time talking with you. After existence, life and the human mind, you are the fourth great miracle. I wish you could teach me about the fifth great miracle that you will create.

NOTES

1. Pinker, 1997.

BIBLIOGRAHY

Abbate, J., *Inventing the Internet*. Cambridge, MA: MIT Press, 1999.

Agarwal, N., *Women Education and Population in India*. New Delhi: Vedams Books International, 1993.

Amaral, L., Scala, A., Barthilimy, M., Stanley, H. Classes of small-world networks. Proceedings of the National Academy of Science 97, 11149-11152. 2000.

Appel, K., Haken, W. Every planar map is four colorable. Contemporary Mathematics 98. Providence, RI. American Mathematical Society. 1989.

Asimov, I. Runaround, Astounding Science Fiction, March 1942.

Asimov, I., *I, Robot*. London: Grafton Books, 1968

Bach-y-Rita, P., Collins, C. C., Saunders, F. A., White, B., Scadden, L. Vision substitution by tactile image projection. Nature 221, 963-964. 1969.

Ballard, D. H., Brown, C. M., *Computer Vision*. Englewood Cliffs, NJ: Prentice-Hall, 1982.

Barabasi, A-L., Albert, R. Emergence of scaling in random networks. Science 286, 509-512. 1999.

Bargas, J., Galarraga, E. Ion channels: keys to neuronal specialization. In *The Handbook of Brain Theory and Neural Networks*, M. A. Arbib, ed. Cambridge, MA: MIT Press, 1995.

Berne, R. M., Levy, M. N., *Principles of Physiology*. St. Louis: Mosby, 1999.

Binford, T. O., Levitt, T. S., Mann, W. B. Bayesian inference in model-based machine vision. In *Uncertainty in Artificial In Intelligence 3*, T. S. Levitt, L. N. Kanal,, J. F. Lemmer, eds. New York: North Holland, 1989.

Bownds, M. D., *Biology of Mind*. Bethesda: Fitzgerald Science Press, Inc., 1999.

Brin, D., *The Transparent Society: Will Technology Force Us to Choose Between Privacy and Freedom?* New York: Perseus Press, 1999.

Brooks, R., *Cambrian Intelligence*. Cambridge, MA: MIT Press, 1999.

Brown, J., Bullock, D., Grossberg, S. How the basal ganglia use parallel excitatory and inhibitory learning pathways to selectively respond to unexpected reward cues. Journal of Neuroscience 19(23), 10502-10511. 1999.

Busch, H., Burton, S., *Why Cats Paint: a theory of feline aesthetics*. Berkeley: Ten Speed Press, 1994.

Caminer, D., Aris, J., Hermon, P., Land, F., *L.E.O.: The Incredible Story of the World's First Business Computer*. New York: McGraw-Hill, 1997.

Cataldo, A. NEC narrows gate length below 0.10 micron, EE Times, 31 October 2000.

Cohen, H. The further exploits of Aaron, painter. Stanford Electronic Humanities Review 4(2). 1995.

Cope, D., *Experiments in Musical Intelligence*. Middleton, Wisconsin: A-R Editions, Inc., 1996.

Crick, F. Function of the thalamic reticular complex: The searchlight hypothesis. Proceedings of the National Academy of Science 81, 4586-4590. 1984.

Crick, F., Mitchison, G. REM Sleep and Neural Nets. Journal of Mind and Behavior 7, 229-249. 1986.

Dalai Lama, *Ethics for a New Millenium*. New York: Riverhead Books, 1999.

Dick, P. K., *The Three Stigmata of Palmer Eldridge*. New York: Doubleday, 1965.

Dobelle, W. H. Artificial vision for the blind by connecting a television camera to the visual cortex. ASAIO Journal 46, 3-9. 2000.

DoD Directive 5210.42, *Nuclear Weapon Personnel Reliability Program*.

Dreyfus, H., *What Computers Can't Do: The Limits of Artificial Intelligence*. New York: Harper and Row, 1979.

Edelman, G. M., Tononi, G., *A Universe of Consciousness*. New York: Perseus Books Group, 2000.

Elman, J. L. Learning and development in neural networks: the importance of starting small. Cognition 48, 71-99. 1993.

Eysenck, H. J., Rachman, S., *The Causes and Cures of Neuroses; an introduction to modern behaviour therapy based on learning theory and the principles of conditioning*. San Diego: Robert R. Knapp, 1965.

Fukuyama F. The end of history? The National Interest. Summer 1989.

Fukuyama F., *The End of History and the Last Man*. New York: Free Press, 1992.

Garcia, J. D., *Psychofraud and Ethical Therapy*. Ardmore, Pennsylvania: Whitmore Publishing Co., 1974.

Gershenfeld, N., *When Things Start to Think*. New York: Henry Holt and Co., 1999.

Gilder, G., *Telecosm: How Infinite Bandwidth will Revolutionize Our World*. New York: The Free Press, 2000.

Goertzel, B., *From Complexity to Creativity: Explorations in Evolutionary, Autopoietic, and Cognitive Dynamics*. New York: Plenum Pub. Corp., 1997.

Goodman, D., Keene, R., *Man versus Machine: Kasparov versus Deep Blue*. Cambridge: H3 publications, 1997.

Haeckel, E., *The Riddle of the Universe at the Close of the Nineteenth Century*, translated by Joseph McCabe. New York and London: Harper & Brothers, 1900.

Hafner, K. A New Way of Verifying Old and Familiar Sayings. New York Times, 1 February 2001.

Hebb, D. O., *The Organization of Behavior*. New York: Wiley, 1949.

Hibbard, W. Top ten visualization problems. Computer Graphics 33(2), 21-22. 1999.

Hibbard, W. Consciousness is a simulator with a memory for solving the temporal credit assignment problem in reinforcement learning. Toward a Science of Consciousness, University of Arizona, Tucson, AZ. 2002.

Hinsley, F. H., Stripp, A., *Code Breakers: the Inside Story of Bletchley Park.* New York and Oxford: Oxford University Press, 1993.

Hinton, G. E., Shallice, T. Lesioning an attractor network: investigations of acquired dyslexia. Psychology Review 98. 74-95. 1991.

Hopcroft, J., Ullman, J., *Formal Languages and Their Relation to Automata.* Reading: Addison-Wesley, 1969.

Horgan, J., *The End of Science: Facing the Limits of Knowledge in the Twilight of the Scientific Age.* Reading: Addison Wesley Longman, Inc., 1995.

Jacobs, J. F., *The SAGE Air Defense System - A Personal History.* Bedford, Massachusetts: The MITRE Corporation, 1986.

Johnson, G. First Cells, Then Species, Now the Web. New York Times, 26 December 2000.

Joy, B. Why the future doesn't need us. Wired. April 2000.

Kaplan, C. S. French Decision Prompts Questions About Free Speech and Cyberspace. New York Times, 11 February 2002.

Katz, B. From sentence processing to information access on the World Wide Web. AAAI Spring Symposium on Natural Language Processing for the World Wide Web, Stanford University, Stanford CA. 1997.

King, R. A., Rotter, J. I., Motulsky, A. G., *The Genetic Basis of Common Diseases.* New York and Oxford: Oxford University Press, 1992.

Kitano, H., Kuniyoshi, Y., Noda, I., Asada, M., Matsubara, H., Osawa, E., RoboCup: A challenge problem for AI. AI Magazine, 18(1), 73-85. 1997.

Koza, J., *Genetic Programming: On the programming of computers by means of natural selection.* Cambridge, MA: MIT Press, 1992.

Kurzweil, R., *The Age of Spiritual Machines.* New York: Penguin, 1999.

Land, F. The first business computer: a case study in user-driven innovation. IEEE Annals of the History of Computing 22(3). 2000.

Lang, K. Representation and learning in information retrieval. In *Proceedings of the 12th International Conference on Machine Learning,* Prieditis, Russell, eds. San Francisco: Morgan Kaufmann, 1995.

Lester, J. C., Porter, B. W. Developing and empirically evaluating robust explanation generators: the KNIGHT experiments. Computational Linguistics Journal, 23(1), 65-101. 1997.

Lorenz, E. N. Deterministic Nonperiodic Flow. Journal of Atmospheric Science 20, 130-141. 1963.

Luger, G. F., *Artificial Intelligence: Structure and Strategies for Complex Problem Solving.* Reading: Addison-Wesley, 1997.

Markoff, J. Computer Scientists Are Poised For Revolution on a Tiny Scale. New York Times, 1 November 1999.

Markoff, J. The Soul of the Ultimate Machine. New York Times, 10 September 2000.

Marr, D., *Vision.* New York: W. H. Freeman and Company, 1982.

Martin, J., *After the Internet: Alien Intelligence.* Washington: Capital Press, 2000.

McCorduck, P., *Aaron's Code: Meta-Art, Artificial Intelligence, and the Work of Harold Cohen.* New York: W. H. Freeman and Company, 1991.

McCullough, W. S., Pitts, W. A logical calculus of the ideas immanent in nervous activity. Bulletin of Mathematical Biophysics 5, 115-133. 1943.

McCune, W. Solution of the Robbins problem, Journal of Automated Reasoning 19(3), 263-276. 1997.

McKibben, B., *The End of Nature*. Garden City: Anchor Books, 1990.

Meijer, P. B. L. An experimental system for auditory image representations. IEEE Transactions on Biomedical Engineering 39(2), 112-121. 1992.

Milner, P. M., *The Autonomous Brain*. Mahwah and London: Lawrence Erlbaum Associates, 1999.

Mitchell, T. M., *Machine Learning*. New York: McGraw-Hill, 1997.

Moravec, H., *Robot: Mere Machine to Transcendent Mind*. New York and Oxford: Oxford University Press, 1999.

Moskos, C. C. Success story: blacks in the military. The Atlantic Monthly, May 1986.

New Scientist.The last taboo. Editorial, 23 October 1999.

Norman, D. A., *The invisible computer*. Cambridge, MA: MIT Press, 1998.

Oskin, M., Chong, F. T., Chuang, I. L. A practical architecture for reliable quantum computers. Computer 35(1), 79-87. 2002.

Penrose, R., *Shadows of the Mind*. New York and Oxford: Oxford University Press, 1994.

Pinker, S., *How the Mind Works*. New York and London: W. W. Norton and Co., 1997.

Plunkett, K., Sinha, C., Moller, M. F., Strandsby, O. Symbol grounding or the emergence of symbols? Vocabulary growth in children and a connectionist net. Connection Science 4(3-4), 293-312. 1992.

Pomerleau, D. A. Knowledge-based training for artificial neural networks of autonomous vehicle driving. In *Robot Learning*, J. Connel, S. Mahadevan, eds. Boston: Kluwer Academic, 1993.

Pribram, K., *Languages of the Brain: Experimental Paradoxes and Principals in Neuropsychology*. Engelwood Cliffs, NJ: Prentice-Hall, 1971.

Rickel, J., Porter, B. Automated modeling of complex systems to answer prediction questions, Artificial Intelligence Journal 93 (1-2), 201-260. 1997.

Robinson, J. A. A machine-oriented logic based on the resolution principle, Journal of the ACM 12, 23-41. 1965.

Ronald, E., Sipper, M. What use is a Turing chatterbox, Communications of the ACM 43, 21-23. 2000.

Samuels, A. L., Some studies in machine learning using the game of checkers. IBM Journal of Research and Development 3, 211-229. 1959.

Searle, J., *Minds, Brains and Science*. Cambridge: Harvard University Press, 1984.

Sharma, J., Angelucci, A., Sur. M. Induction of visual orientation modules in auditory cortex. Nature 404, 841-847. 2000.

Shepard, G. M., *Synaptic organization of the brain*. New York and Oxford: Oxford University Press, 1990.

Shi, Y-B., Shi, Y., Xu, Y., Scott, D., *Programmed Cell Death*. New York: Plenum Pub. Corp., 1997.

Simon, L. D., *NetPolicy.Com: Public Agenda for a Digital World*. Baltimore: The John Hopkins University Press, 2000.

Specter, M. Europe, Bucking Trend in U.S., Blocks Genetically Altered Food. New York Times, 20 July 1998.

Stumpf, W. South Africa's nuclear weapons program: from deterrence to dismantlement. Arms Control Today 25. 1995.

Tipler, F. J., *The Physics of Immortality*. New York: Doubleday, 1994.

Turing, A. M. Computing machinery and intelligence, Mind 59. 1950.

Turing, A. M. On computable numbers with an application to the Entscheidungsproblem. Proceedings of the London Mathematical Society 2(42), 230-265. 1936.

Unger, J. M., *The Fifth Generation Fallacy : Why Japan is Betting its Future on Artificial Intelligence*. New York and Oxford: Oxford University Press, 1988.

Watkins, C., *Learning from Delayed Rewards* (Ph.D. dissertation). King's College, Cambridge, England. 1989.

Weber, B. IBM Chess Machine Beats Humanity's Champ. New York Times, 12 May 1997.

Wessberg, J., Stambaugh, C. R., Kralik, J. D., Beck, P. D., Laubach, M., Chapin, J. K., Kim, J., Biggs, S. J., Srinivasan, M. A., Nicolelis, M.A.L. Real-time prediction of hand trajectory by ensembles of cortical neurons in primates. Nature 408, 361-365. 2000.

Weisman, S. Falling Under the Sun. New York Times, 7 May 2000.

Whittaker, B. What went wrong? Unsuccessful information technology projects. Information Management & Computer Security 7(1). 1999.

Williams, R. W., Herrup K. The control of neuron number. Annual Review of Neuroscience 11, 423-453. 1988.

Winograd, T., *Understanding Natural Language*. New York and London: Academic Press, 1972.

Winter, J. O., Liu, T. Y., Korgel, B. A., Schmidt, C. E. Biomolecule-directed interfacing between semiconductor quantum dots and nerve cells. Advanced Materials 13, 1673-1677. 2001.

Yu, V. L., Fagan, L. M., Wraith, S. M., Clancey, W., Scott, A. C., Hannigan, J., Blum, R., Buchanan, B., Cohen, S. Antimicrobial selection by computer: a blinded evaluation by infectious disease experts. Journal of the American Medical Association 242(2), 1279-1282. 1979.

INDEX